# The Use of Ropes

Homer J. Dana, W. A. Pearl

Alpha Editions

This edition published in 2024

ISBN : 9789362093332

Design and Setting By
**Alpha Editions**
www.alphaedis.com
Email - info@alphaedis.com

As per information held with us this book is in Public Domain.
This book is a reproduction of an important historical work. Alpha Editions uses the best technology to reproduce historical work in the same manner it was first published to preserve its original nature. Any marks or number seen are left intentionally to preserve its true form.

# Contents

INTRODUCTION ................................................................- 1 -
KNOTS ................................................................- 2 -
SPLICES ................................................................- 17 -
HITCHES ................................................................- 21 -
LASHINGS ................................................................- 37 -
TACKLE SETS ................................................................- 39 -
CHAIN HOISTS ................................................................- 48 -
TRANSMISSION CABLES ................................................................- 51 -
STEEL CABLES ................................................................- 54 -
SPLICING TRANSMISSION CABLES ................................................................- 65 -

# INTRODUCTION

Each year, old industries keep expanding and new ones are created. In many of these, the use of hoists, tackle, rope transmissions, etc. is ever increasing in extent and importance. Information on the selection and use of ropes and tackles and the tying of knots is very scattering and incomplete. The purpose of this bulletin is to collect information from all the different sources possible and assemble it under one cover, in the hope that it may be valuable to people in many different fields of activity. It is not meant to be an advanced treatise for those who consider themselves already proficient in the use of rope and tackle but is designed as an aid and reference to those less skilled in the art.

A variety of knots and splices are shown with occasional suggestions as to their use and application. Some knots tie easily and are very secure but are not so easy to untie; others are easily and quickly tied—are secure and yet are not difficult to untie. Some knots are suitable for small cords only, and others are adapted to large ship's hawsers. For these and other reasons, it is desirable to select the right knot for the job in hand.

Nearly every individual at some time or other has gone camping. If he chanced to select a remote or inaccessible mountain side for a vacation trip, he probably had one or more pack animals to take in the supplies and camp outfit. How many could use the famous Diamond hitch to fasten the pack on the horse's back so that it will not shift or fall off in transit?

The desirability of correct selection with reference to the work to be done is also true of tackle sets. One type of tackle will give great mechanical advantage, but requires an excessive amount of rope or requires frequent overhauling to complete the job, while another type, using the same equipment, will not give such great mechanical advantage but does not require overhauling so often during the progress of the load.

Rope is coming more and more into favor for the transmission of power—replacing gears and heavy leather belts. It is important that the proper sized sheave wheel be used with a rope of given diameter in order to secure the longest service from the transmission. It is also important that speed be considered in the calculation for necessary strength to transmit a certain amount of power. It is evident from these two instances alone that it is desirable that the selection of a rope transmission should be governed by the use of complete sets of data on the subject.

Some of the knots, splices, etc. shown in this bulletin were found to have more than one name, or were called by different names by different authors. In such case only the most commonly used term was selected.

# KNOTS

A knowledge of knots has saved many a life in storm and wreck, and if everyone knew how to tie a knot quickly and securely there would be fewer casualties in hotel and similar fires where a false knot in the fire escape rope has slipped at the critical moment and plunged the victim to the ground. Many an accident has occurred through a knot or splice being improperly formed. Even in tying or roping a trunk, few people tie a knot that is secure and quickly made and yet readily undone. How many can tie a tow rope to a car so it will be secure and yet is easily untied after the car has been hauled out of the mud? Or suppose a rope was under strain holding a large timber in midair and a strand in the derrick guy rope shows signs of parting. How many could attach a rope each side of the weak spot to take the strain?

The principle of a knot is that no two parts which lie adjacent shall travel in the same direction if the knot should slip. Knots are employed for several purposes, such as, to attach two rope ends together, to form an enlarged end on a rope, to shorten a rope without cutting it, or to attach a rope to another rope or object. Desirable features of knots are that they may be quickly tied, easily untied and will not slip under a strain. In a number of cases a toggle is used either to aid in making the knot or make it easier to untie after a strain has been applied.

A number of terms are commonly used in tying knots. The "standing" part is the principal portion, or longest part of the rope. The "bight" is the part curved, looped or bent, while working or handling the rope in making a knot, and the "end" is that part used in forming the knot or hitch. The loose, or free end, of a rope should be knotted or whipped to prevent it from raveling while in use.

## Strength of Knots

If a knot or hitch of any kind is tied in a rope its failure under stress is sure to occur at that place. Each fiber in the straight part of the rope takes its proper share of the load, but in all knots the rope is cramped or has a short bend, which throws an overload on those fibers that are on the outside of the bend and one fiber after another breaks until the rope is torn apart. The shorter the bend in the standing rope the weaker the knot. The approximate strength of several types of knots in percent, of full strength of a rope is given in the table below, as an average of four tests.

1. Full strength of dry rope          100%

2. Eye splice over an iron thimble     90%

3. Short splice in rope                          80%

4. Timber hitch, round turn and half hitch       65%

5. Bowline, slip knot, clove hitch               60%

6. Square knot, weaver's knot, sheet bend        50%

7. Flemish loop, over—hand knot                  45%

### Fastening Knots

<u>Fig. 1.</u> The over-hand knot is the simplest of all knots to make. It is made by passing the lose end of the rope over the standing part and back through the loop.

<u>Fig. 2.</u> The Double knot is made by passing the free end of the rope through the loop twice instead of but once as in making an over-hand knot. This is used for shortening or for a stop on a rope, and is more easily untied than the over-hand knot. It is also known as a blood knot, from its use on whip lashes by slave drivers, etc.

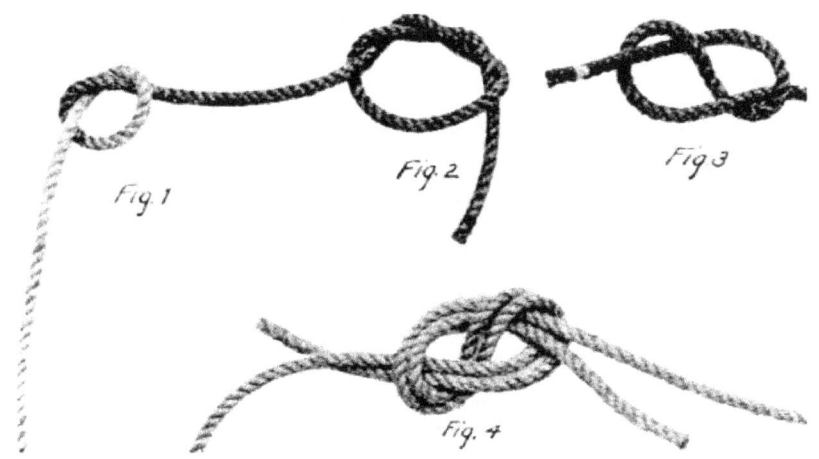

<u>Fig. 3.</u> The Figure Eight knot is similar to the over-hand knot except that the loose end of the rope is passed through the loop from the opposite side. It is commonly used to prevent a rope running through an eye or ring or tackle block. It is also used as the basis for ornamental knots, etc.

<u>Fig. 4.</u> The Double Figure Eight knot is made by forming a regular figure eight and then following around with the end of the other rope as shown.

<u>Fig. 5.</u> The Square knot is probably the commonest and most useful of all knots. It is strong and does not become jammed when being strained. Take

the ends of the two ropes and pass the left end over and under the right end, then the right end over and under the left. Beware of the granny knot which is often mistaken for the square knot but is sure to slip under strain.

Fig. 6. The Reef knot is a slight modification of the square knot. It consists merely of using the bight of the left or right end instead of the end itself, and is tied exactly as is the square knot. This makes the knot easy to untie by pulling the free end of the bight or loop.

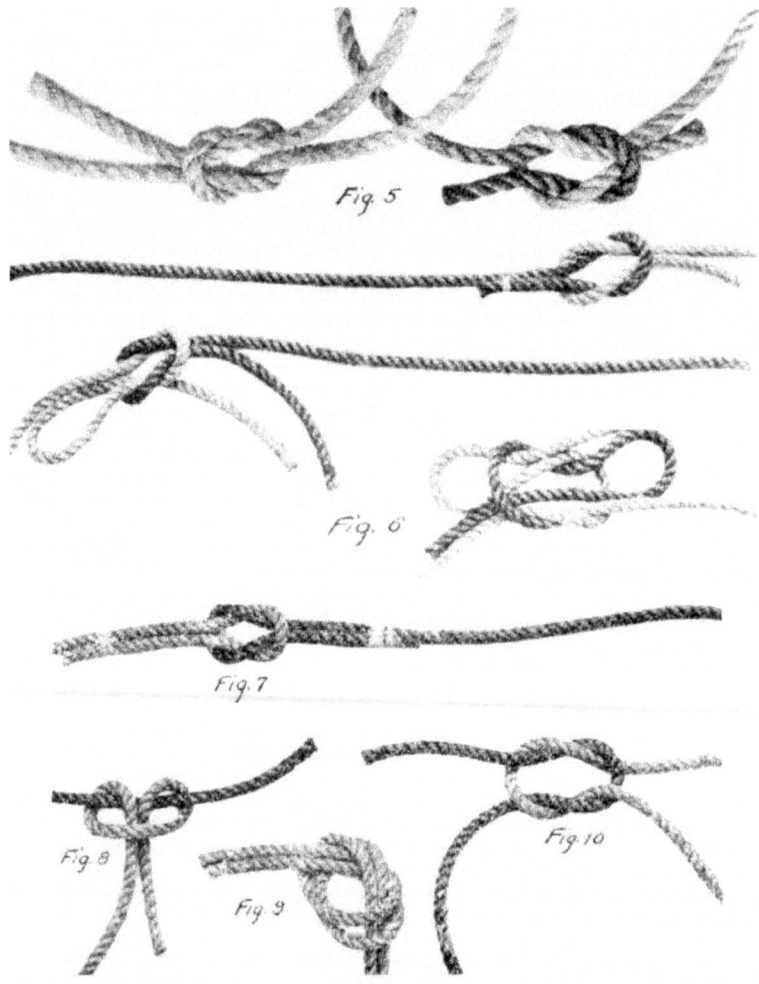

Fig. 7. If the Square or reef knot is used to join two ropes of unequal diameter, the knot is apt to slip unless the ends of the rope are whipped as shown.

- 4 -

Fig. 8. A Square knot joining two ropes of unequal size is apt to slip with a result similar to that shown.

Fig. 9. The Open-hand knot is made by tying an over-hand knot with two rope ends lying parallel. It is better than a square knot for joining two ropes of unequal diameter. Grain binders use this knot.

Fig. 10. The Granny knot is often mistaken for a square knot and its use should by all means be avoided as it is almost sure to slip when a strain is applied, unless the ends are whipped. For large rope, a granny knot with ends whipped will hold securely and is easy to untie.

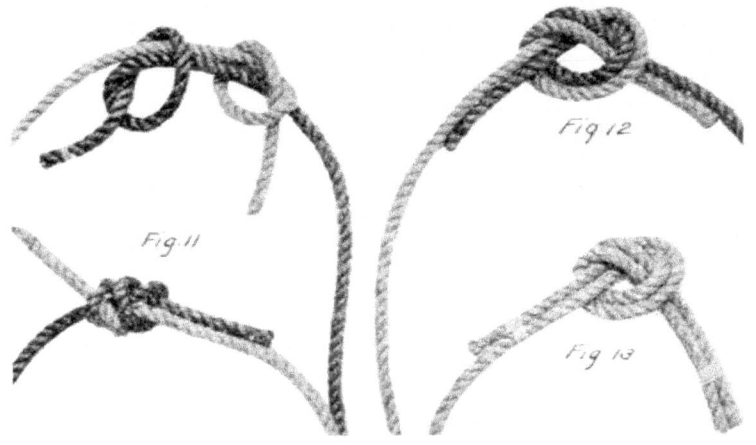

Fig. 11. The Fisherman's knot is a simple type of knot formed by two simple over-hand knots slipped over the standing parts of the two ropes, and drawn tight. It is valuable for anglers as the two lines may be drawn apart by merely pulling on the loose ends of the rope.

Fig. 12. The Ordinary knot is used for fastening two heavy ropes together and is made by forming a simple knot with the end of one rope and then interlacing the other rope around it, as shown.

Fig. 13. Whipping the two ends of an Ordinary knot makes it more secure.

Fig. 14. The Weaver's knot is used to join small lines or threads and is made by forming a bight in one rope, passing the end of the second rope around the bight, back over itself and through the bight. Weavers use this knot in tying broken threads. When pulled tight, both ends point backward, and do not catch when pulled thru the loom.

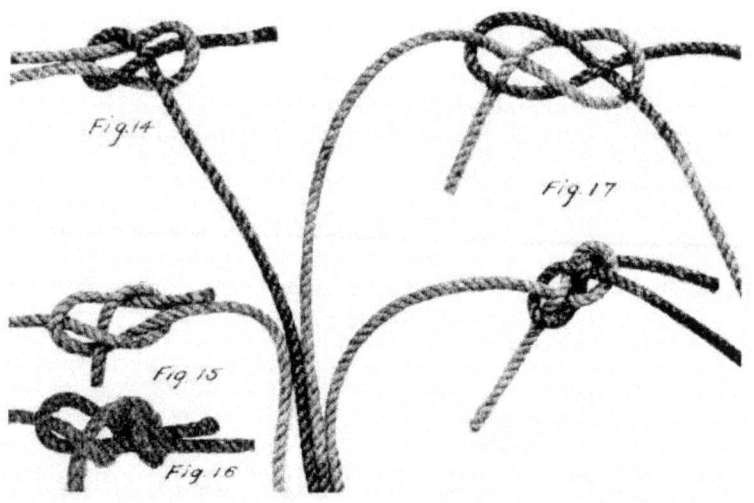

Fig. 15. The Hawser knot or sheet bend is used for joining stiff or heavy ropes and is not to be confused with the weaver's knot. It resembles the bowline, and is easily untied.

Fig. 16. The Double Sheet Bend is similar to the Hawser knot and is useful for the same purposes.

Fig. 17. The Garrick bend is commonly used for joining two heavy hawsers which are too stiff to bend easily.

Fig. 18. Another method of joining stiff hawsers is to use the Half-hitch and whipping. This is a satisfactory method of making a joint to be used for a considerable time.

Fig. 19. The Slip knot as shown is a knot with many uses.

Fig. 20. The Bowline knot is useful for forming a loop on the end of a rope. It is used frequently by stockmen to tie a horse or cow so that they will not choke themselves. It is always secure and easily untied. Use this knot in tying a tow rope to a car.

Fig. 21. The Running Bowline is used for the same purposes as the slip knot in Fig. 19, but is much more secure. It will always run freely on the standing part of the rope, and is easily untied.

Fig. 22. A Loop knot is useful for making fast to the middle of a rope where the ends are not free. It will pull tight under strain, and is not easily untied.

Fig. 23. The Tom-fool knot is formed in the middle of a rope and may be used for the same purpose as the loop knot, except in this case either standing part of the rope may be strained without the knot failing, or slipping. It can be used for holding hogs. Place one loop over the hog's snout and hold onto one rope. Release by pulling other rope. Can also be used from the ground for releasing hoisting tackle which has been used on a flag pole or other tall object.

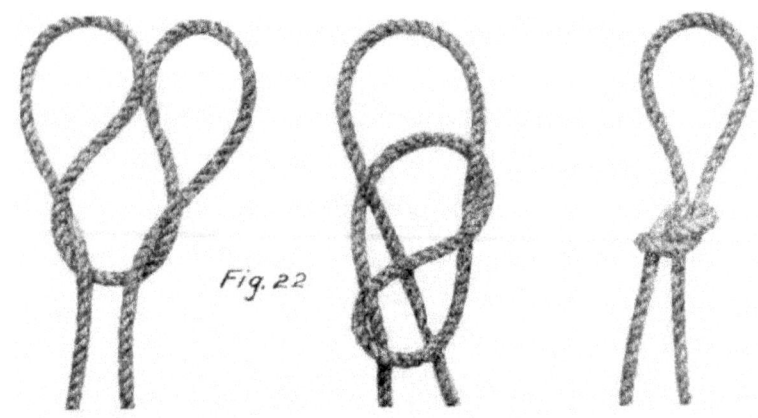

Fig. 24. The Boat knot is formed by the aid of a toggle on a rope whose ends are not free, and is used for shortening or for stopping a ring on a taut line.

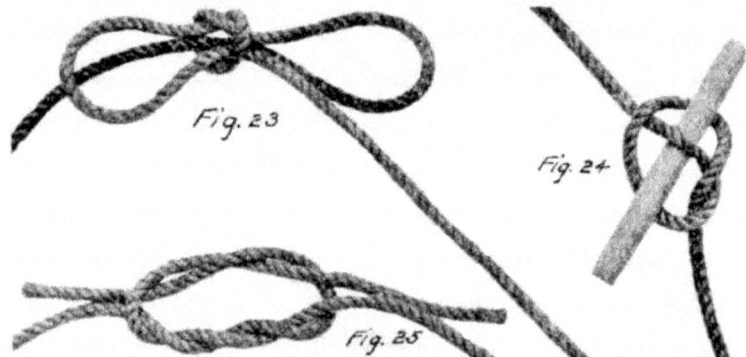

Fig. 25. The Surgeon's knot is a modified form of the square knot, and if used with smooth cord, as in tying bundles, it holds very securely. The object of the double twist is to make the knot easy to tie without holding with the end of the finger.

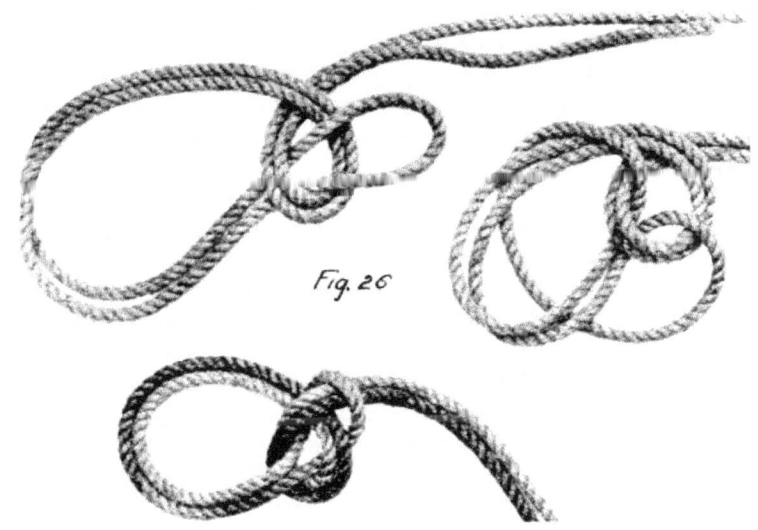

Fig. 26. Bowline on the bight is easily made on the looped part of a rope which is double. It is used where a loop is desired which will not pull tight or choke and is easily untied. May be used for casting harness for horses.

Fig. 27. The Spanish Bowline is a knot which may be made in the middle of a long rope or in a bight at the end, and gives two single loops that may be thrown over two separate posts or both thrown over one. Either loop will hold without slipping and is easily untied.

Fig. 28. The Flemish loop is similar to the Fisherman's knot, Fig. 11, except that it is used for forming a loop on the end of a rope instead of joining two ropes. The loop or eye will not close up when strained.

Fig. 29. The Hawser knot with toggle is formed exactly the same as the regular Hawser knot except that the toggle is inserted for the purpose of making it easy to loosen the knot after a strain has been applied.

**Ending Knots**

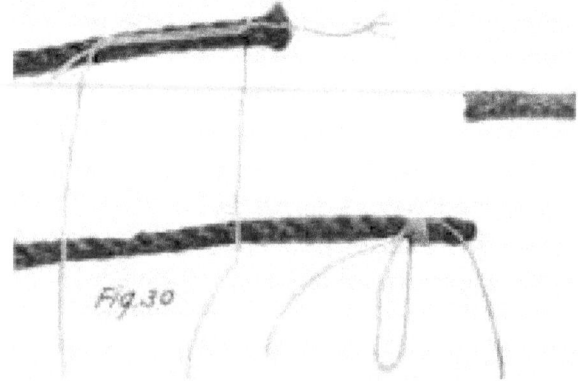

A group of knots somewhat different from those already described, are those used for ending ropes. Ending knots not only serve the purpose of giving a large end on the rope, but also take the place of whipping, in that they prevent the rope from unraveling. Sometimes an ending knot is also used for its ornamental value.

Fig. 30. A Whipping applied as shown is employed for keeping loose ends from fraying or unraveling, where the use to which the rope is to be put will not permit of a knot on the end. Strong cord is used for whipping. In splicing ropes, the whipping is removed before the splice may be considered complete.

Fig. 31. The Single Crown, tucked, makes the rope end but slightly larger than the standing part, and serves to prevent the strands from unraveling. This gives a neat appearing end. To make this type of knot, leave the ends long enough so they can be brought down and tucked under the strands of the standing part. After tucking them under the first strand, as shown, halve each strand and tuck it again under the next strand of the standing part and continue this until the ends are completely tucked the whole length, thus giving a gradual taper to the end of the rope and also giving a knot that will stand by itself. The single crown not tucked, is not a good ending for a rope.

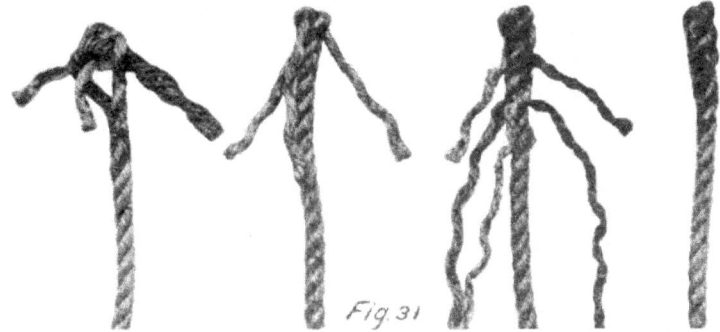

Fig. 31

Fig. 32. The Wall knot is frequently used as an ending knot to prevent unraveling. It is very satisfactory where the rope does not need to pass through a block or hole which is but slightly larger than itself. The Wall knot may be tucked similar to the Crown and makes a very secure ending for a rope. For small ropes unlay the strands back, each three inches, and on larger ropes in proportion. Hold the rope in the left hand with the loose strands upward. With the right hand take the end of strand number one and bring it across the loose end in position with the thumb of the left hand, then take the rope, forming a loop and allowing the end to hang free. Hold strand number two and pass it under strand number one and hold it against the rope with the thumb of the left hand. Again with the right hand take strand number three and pass it under strand number two and up through the first loop formed. Then draw each of the strands gradually until the knot is tightened.

Fig. 33. The Matthew Walker knot or Stopper knot is similar to the Wall knot except the ends are inserted through two loops instead of one as in the Wall knot. It can readily be made by loosely constructing the Wall knot as explained before and continuing as follows: pass the end number one through the loop with two, then end number two through the loop with three, and number three through the loop with one, then gradually tighten the knot by drawing in a little at a time on each strand.

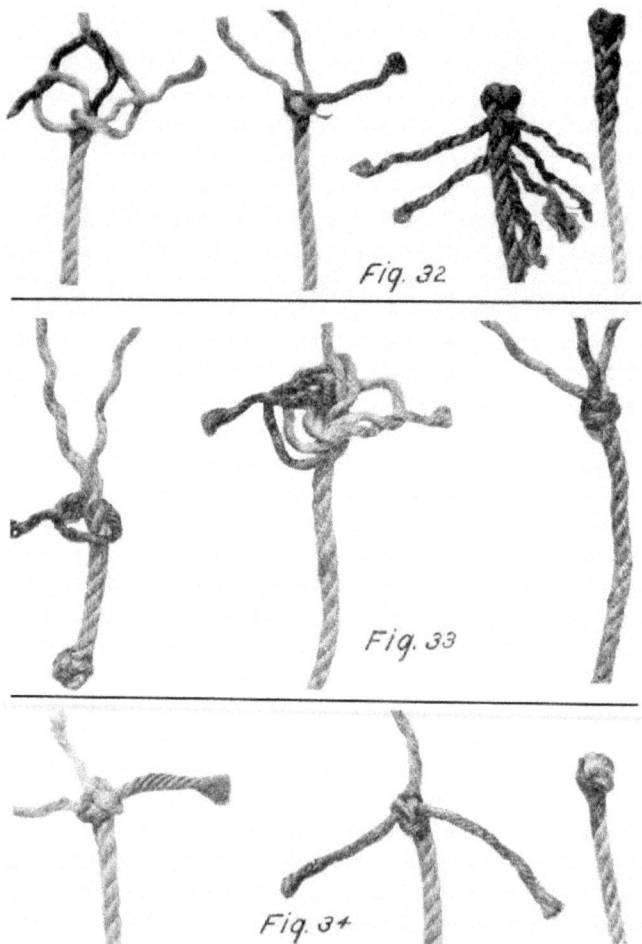

Fig. 32

Fig. 33

Fig. 34

Fig. 34. The Double Wall or Crown knot is made exactly the same as the Single Crown or Wall knot, but instead of trimming off or tucking the ends in, they are carried around a second time, following the lay of the first as shown, and then the knot is pulled tight. When completed, the ends may be tucked in as was done in the Single Crown, or they may be trimmed off.

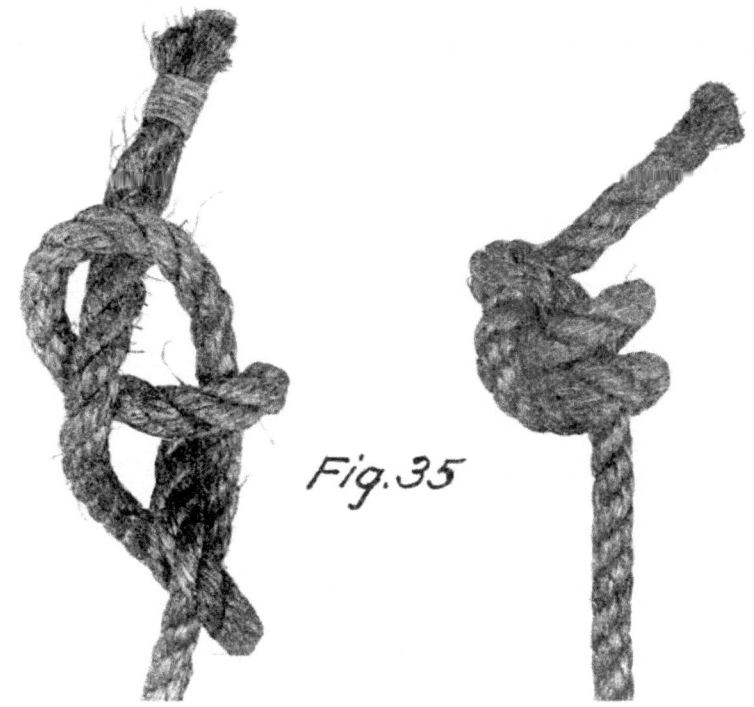

Fig. 35. The Stevedore knot is similar to the Over-hand knot shown in Fig. 1, except that the end of the rope is served around the standing part two and half times before it is tucked through the bight. It is used where a knot is desired to keep the rope from running through a block or hole.

### Shortening Knots

A third type of knots are those which are used where a rope is too long and where it is awkward to have the free ends hanging loose or where the ends are in use and the slack must be taken up in the middle of the rope. These are known as shortening knots. They are also sometimes used merely for ornament.

Fig. 36. The Chain knot is frequently used for shortening and is made by forming a running loop, then drawing a bight of the rope through the loop, and a second bight through the first and so on until the rope has been shortened sufficiently. The free end should then be fastened by passing a toggle or the end of the rope through this last loop. To undo this shortening is very simple as all that is necessary is to either remove the toggle from the last loop or remove the end of the rope if it were used, and then pull on the free end until the knot is completely unraveled.

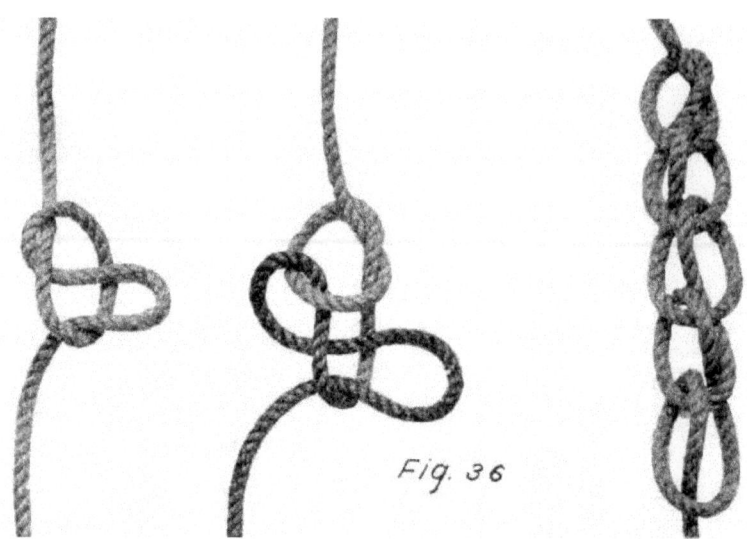

Fig. 36

Fig. 37. The Whipped Shortening or Bend Shortening is one of the most easily made and is well adapted to heavy ropes where a shortening must be made quickly and where it is not to withstand a heavy strain.

Fig. 38. Three-fold Shortening is started by making an Over-hand knot and continuing to tuck the end through the loop three more times, and drawing tight.

Fig. 39. The Sheep-shank or Dog-shank as it is sometimes called, is one of the most widely used of all shortenings. It is made in several forms but the first form shown, while adaptable to fairly stiff ropes, will not withstand much strain. It is used for shortening electric light cords.

Fig. 40. Sheep-shank for free end rope is similar to the plain Sheep-shank except the free end of the rope is passed through the loop. This makes a secure shortening, but it can not be used where the ends of the rope are not free.

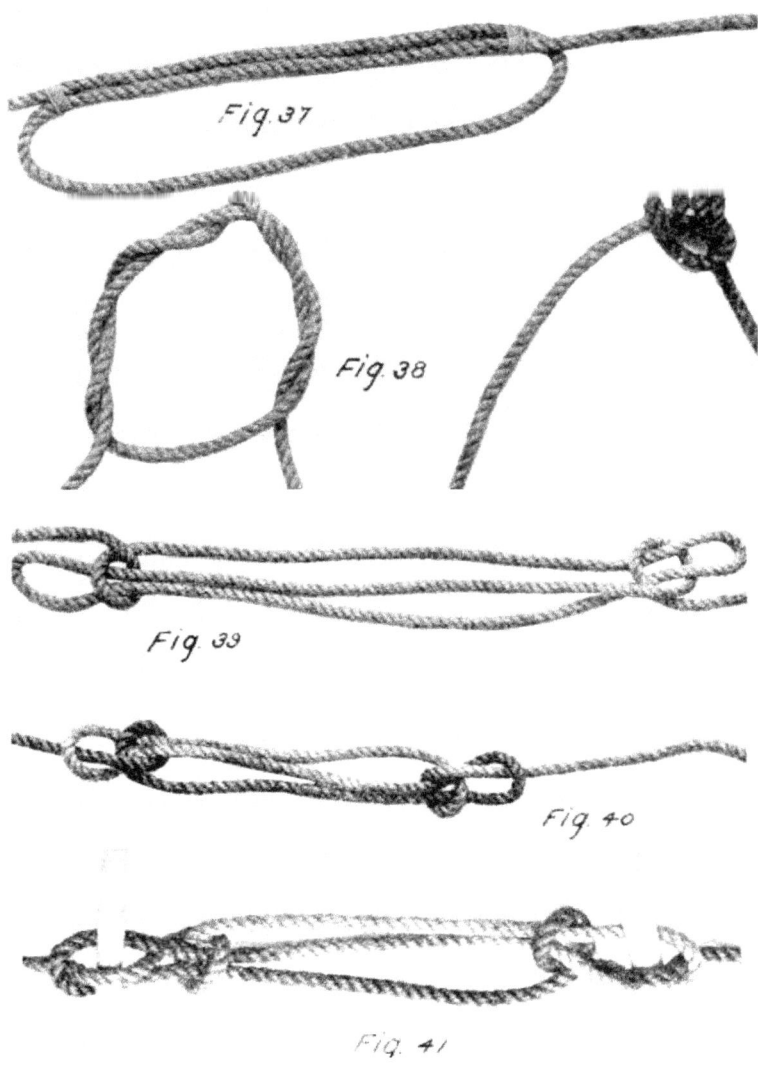

Fig. 41. A Sheep-shank with toggle, is a plain Sheep-shank with the toggle inserted as shown, and makes the shortening as secure as that shown in Fig. 38. It is also easily untied.

Fig. 42.

Fig. 43.

Fig. 42. Sheep-shank with ends whipped is the same as in a plain Sheep-shank except the loop is whipped to the standing part of the rope. This makes the shortening as secure as those shown in Fig. 38, and Fig. 39.

Fig. 43. Bow-shortening is an ordinary knot in the middle of a rope in which a double bend has previously been made. It is not adapted to heavy ropes nor will it stand a heavy strain successfully.

# SPLICES

In the use of ropes, occasion arises, many times, where it is necessary to join two ends together in such a way that the union is as strong as the rest of the rope and still not too large or irregular to pass through a hole or pulley block. Knots are unsuitable in that they will not pass through a block; they are unsightly, and usually are not as strong as the rest of the rope. The method of joining ropes to meet the above requirements is called splicing. There are two general types of rope splices known as the short splice and the long splice. Other applications of the former are made in the eye splice and the cut splice. The long splice is almost always used in splicing wire rope which runs through a block or over a sheave.

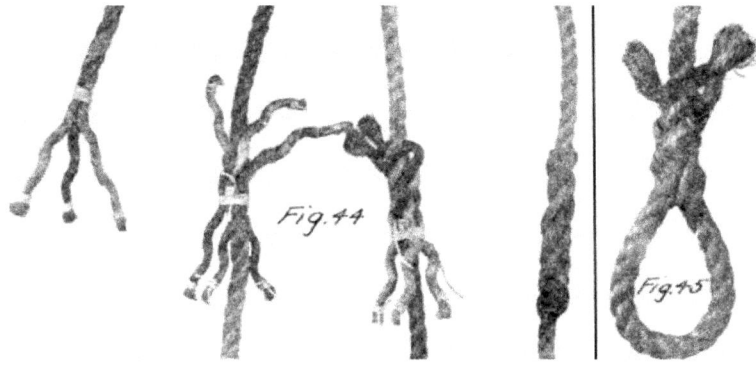

Fig. 44. The Short-splice is made as follows: the two ends to be joined are untwisted for a few inches and the rope is whipped temporarily to prevent further unwinding. The end of each strand is also whipped temporarily to prevent unraveling. The strands may then be waxed if desired. The two rope ends are then locked together or "married" so that the strands from one end pass alternately between those from the other end. The strands from opposite sides are now in pairs. Take two strands from opposite sides as strands A and 1, tie a simple over-hand knot in its right hand form. Similarly with a right hand knot tie together the strands forming each of the pairs B and 2 and C and 3. Draw the knots tight, then pass each strand of the rope over the strand adjacent to it and under the next, coming out between two strands as at first. Repeat until the ends of the strands have been reached—leaving from half an inch to an inch and a half of ends hanging free so that when the rope is put under repeated strain for the first few days, the stretching of the splice will not pull the ends from under the last strand under which they were tucked. After a few days service the free ends may be safely trimmed even with the face of the rope. After the splice has been completed by tucking the ends as above, remove whipping on strands and lay the splice

on the floor and roll it under the foot, or in the case of a large rope, pound it with a mallet to make it round and smooth. The appearance of the splice is improved if the strands are divided in half just before the last tuck is made, and one-half is cut off while the other half is used to complete the splice. This splice may also be made by simply laying the ropes together and then tucking them as above without first tying the simple Over-hand knots. A skilled workman frequently dispenses with the whipping in making a splice.

Fig. 45. An Eye-splice is so much smaller and neater than a knotted eye in the end of a rope that it is much to be preferred to the latter. The Eye-splice is made similar to the short-splice except that the strands on the end of the rope are unlaid for the full length of the splice. The ends are tucked under, over and under, etc., the strands of the standing part of the rope. Stretch well and cut off the loose ends of the strands.

Fig. 46. Long splice. If it is desired to unite two rope ends so that the splices will pass through a pulley as readily and smoothly as the rope itself, what is known as a Long splice is used. This is best suited as it does not cause an enlargement in the rope at the point where the splice is made. To make it, unlay the ends of two ropes to the length of at least five and a half times the circumference of the rope. Interlace the strands as for the Short splice. Unlay one strand and fill up the vacant space which it leaves with the strand next to it from the other rope end. Then turn the rope over and lay hold of the two next strands that will come opposite their respective lays. Unlay one, filling up the vacant space as before, with the other. Take one third out of each strand, knot the opposite one-thirds together and heave them well in place. Tuck all six ends once under adjacent strands and having stretched the splice well, cut off the ends. The ending of successive pairs should occur at intervals in the splice as shown, and not at the two ends as in the Short splice.

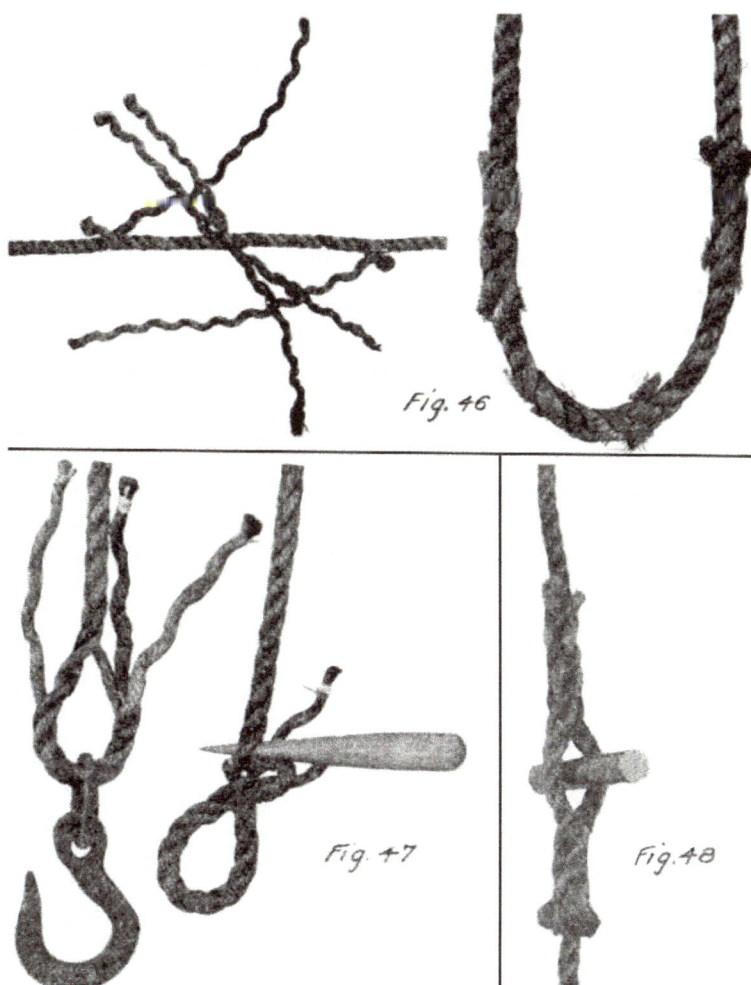

Fig. 47. A Chain splice is used for splicing a rope into a chain end which is required to travel through a block or small opening. It is also sometimes used for making an ordinary eye in the end of a rope. Four or six strand rope lends itself more readily to this type of splice than does a three strand rope. To make a chain splice, unlay the strands more than for an eye splice, then unlay a little further one strand in a three strand rope, and two strands in a four strand rope. Bend the two parts together and tie an Over-hand knot so that the divided strands will lay together again. Continue to lay the ends in by passing them through the eye. When the eye has been completely laid up, the remaining ends should be tucked in the standing part of the rope as in a very short splice. This makes an eye which will not pull out even if the ends of the strands are only whipped without first tucking. It is especially valuable in forming smooth eyes in steel cable, without the use of clamps. In this case,

however, the eye must be made considerably longer than in the case of hemp rope.

Fig. 48. The Cut splice is formed similar to the Eye splice, except that the two rope ends are extended past each other and joined into the standing part of the ropes. This type of splice is frequently used to hold the rings in rope ladders. It can also be used where it is desired to attach a spar or rod to the middle of a line.

# HITCHES

The knots so far described are used mainly for fastening rope ends together or for ending a rope. A quite different class of knots is that used for fastening a rope to a stationary or solid object. This type of knots is known as hitches.

Hitches as well as other types of knots should be easily made, should not slip under strains and should be easily untied. If all ropes were the same size and stiffness it would be possible to select two or three knots which would meet all requirements. But, since this is not true and since a knot suitable for a silken fish line will not be satisfactory for a ship's hawser, we find a great variety of knots, each of which is designed to meet some special requirements of service. The following illustrations show a variety of the most typical and useful knots used on fiber or manila rope.

Fig. 49. The Half-hitch is good only for temporary fastenings where pull is continuous. It is usually used as part of a more elaborate hitch.

Fig. 50. The Timber-hitch is very similar to the Half-hitch but is much more permanent and secure. Instead of the end being passed under the standing part once it is wound around the standing part three or four times as shown.

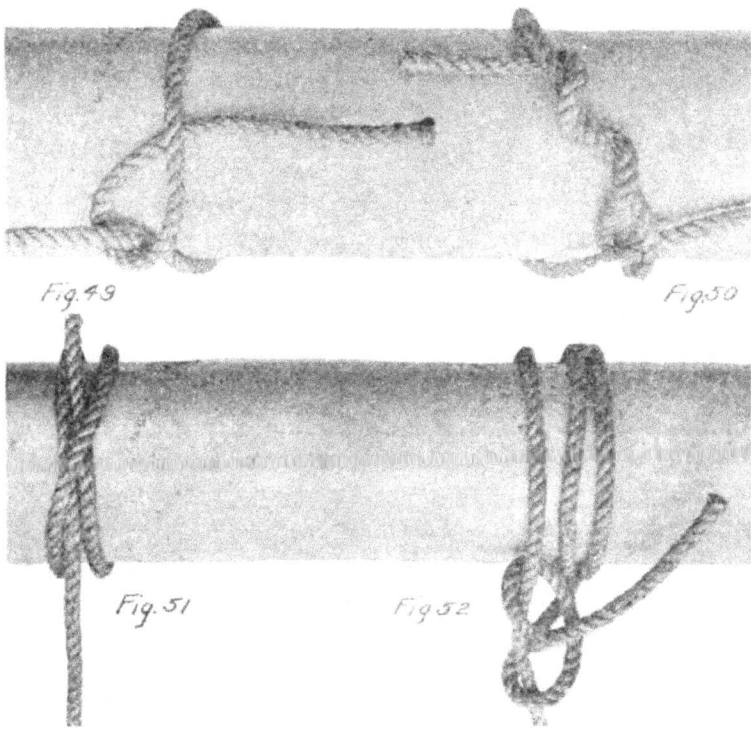

Fig. 51. The Clove, or Builder's-hitch, is more secure than either of the above hitches. It will hold fast on a smooth timber and is used extensively by builders for fastening the staging to upright posts. It will hold without slipping on wet timber. It is also used to make the scaffold hitch.

Fig. 52. The Rolling-hitch is made by wrapping the rope three or four times around the object to which it is to be fastened and then making two half-hitches around the standing part of the rope. It is then drawn tight. This hitch is easily and quickly made and is very secure.

Fig. 53. This illustrates another type of Rolling-hitch very similar to the above but which is not as secure under a heavy strain.

Fig. 54. The Snubbing-hitch is made by passing the rope around the object to which it is desired to fasten it, and then making what is known as a Taut-line hitch, Figure 68, about the standing part of the rope.

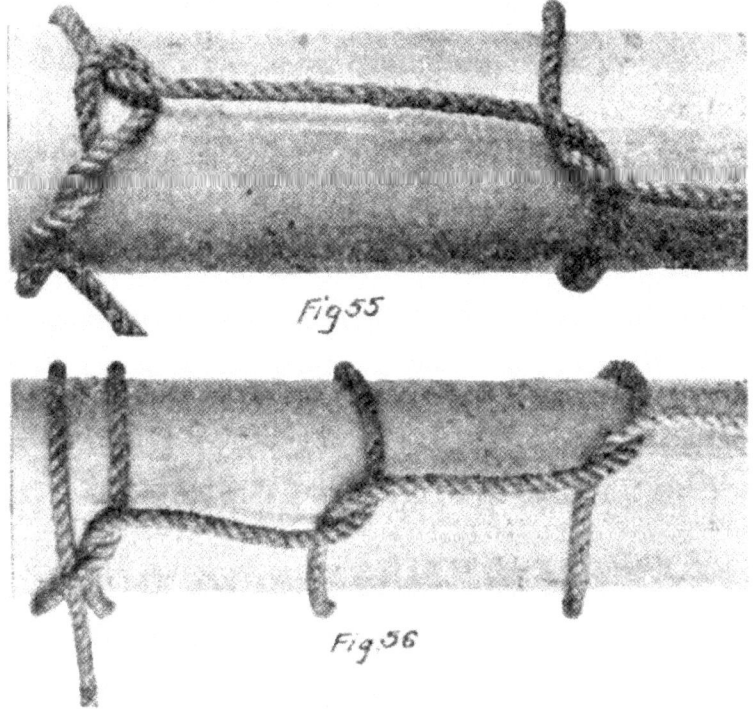

Fig. 55. Timber-hitch and Half-hitch is a combination of the two separate hitches shown in Fig. 49 and Fig 50. It is more secure than either used alone.

Fig. 56. The Chain-hitch is a combination of the above hitch and two or more half-hitches. It is used for hauling in a larger rope or cable with a tow line, etc.

Fig. 57. The Twist-hitch is more secure than the Half-hitch and it is suitable only where the strain is continuous.

Fig. 58. Twist-and-bow-hitch is similar to the Simple Twist-hitch but is easier to untie.

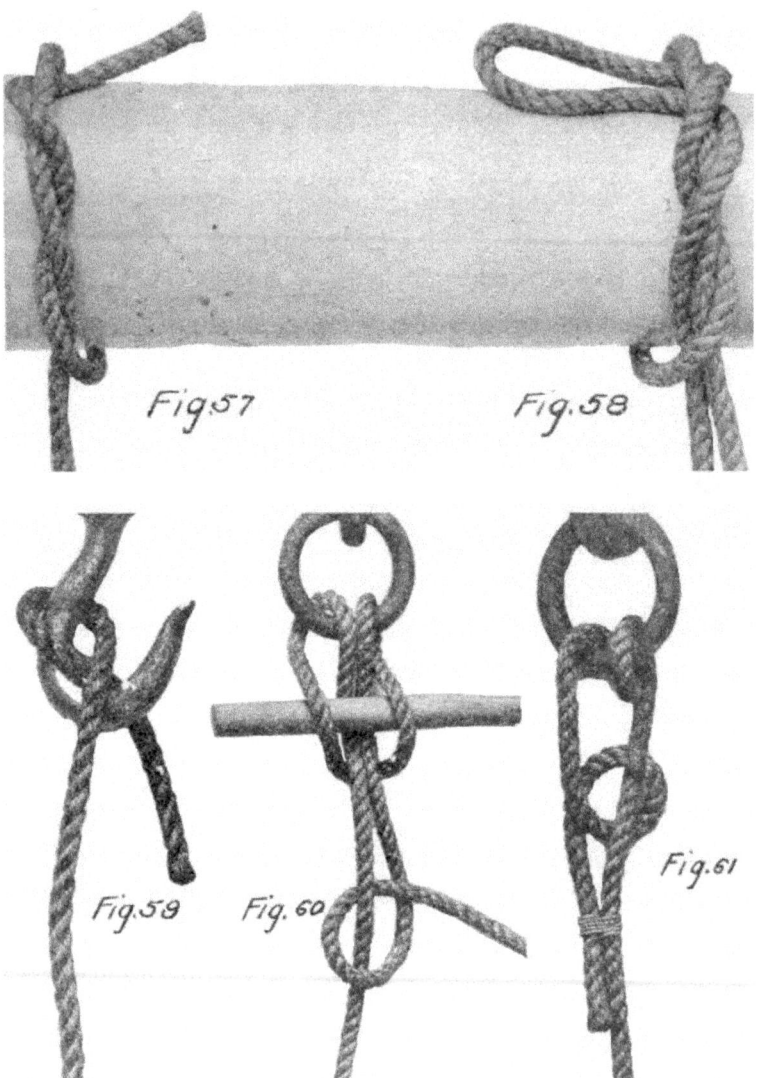

Fig. 59. The Blackwall-hitch is widely used as illustrated. The greater the strain the more securely it holds, but it is unreliable if the rope is slack. This hitch can be used with chain as well as rope.

Fig. 60. The Lark's-head with toggle is easily made and is used as a rule where it is desired to have a type of hitch which is easily and quickly released.

Fig. 61. Round-turn-and-half-hitch is suitable for a more or less permanent method of attaching a rope to a ring. Whipping the end to the standing part of the rope makes it quite permanent.

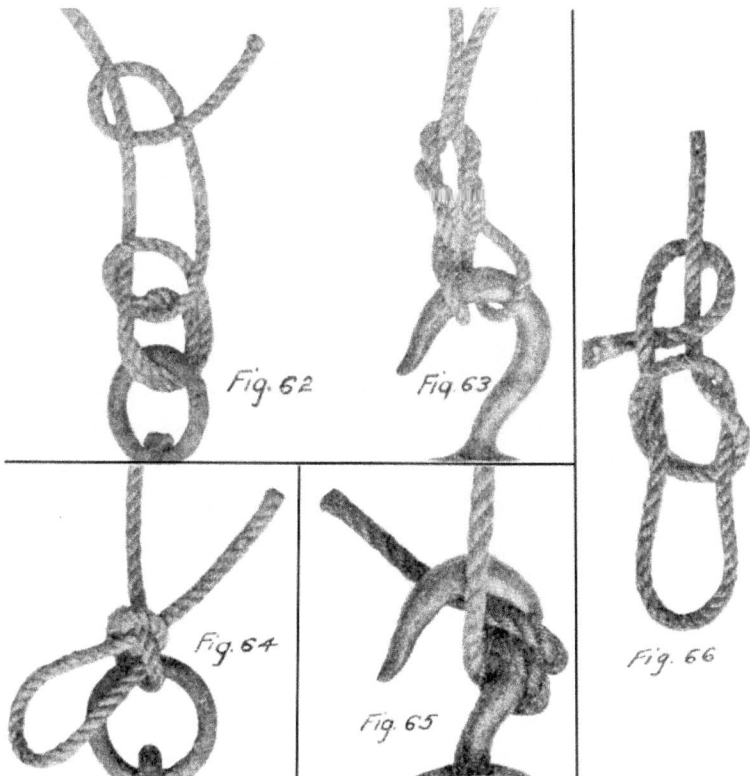

Fig. 62. The Fisherman's hitch is used for fastening large ropes or lines to rings and is very similar to the hitch shown in Fig. 61. It is improved by whipping the free end to the standing part.

Fig. 63. The Cat's-paw-hitch is suitable for attaching a hook to the middle part of a rope where the ends are not free. Strain may be taken on either or both ends. It is easily released.

Fig. 64. The Slippery-hitch is easily made, but has the objection that it draws very tight under strain, making it hard to untie.

Fig. 65. The Double Blackwall is similar to the Single Blackwall and is used for the same purpose.

Fig. 66. The Slip Knot and Half-hitch constitute a combination that is used for the same purpose as the Flemish loop. It is made by first tying a slip knot so that it will run on the short end of the rope. Then complete by tying a half hitch with the short end as shown.

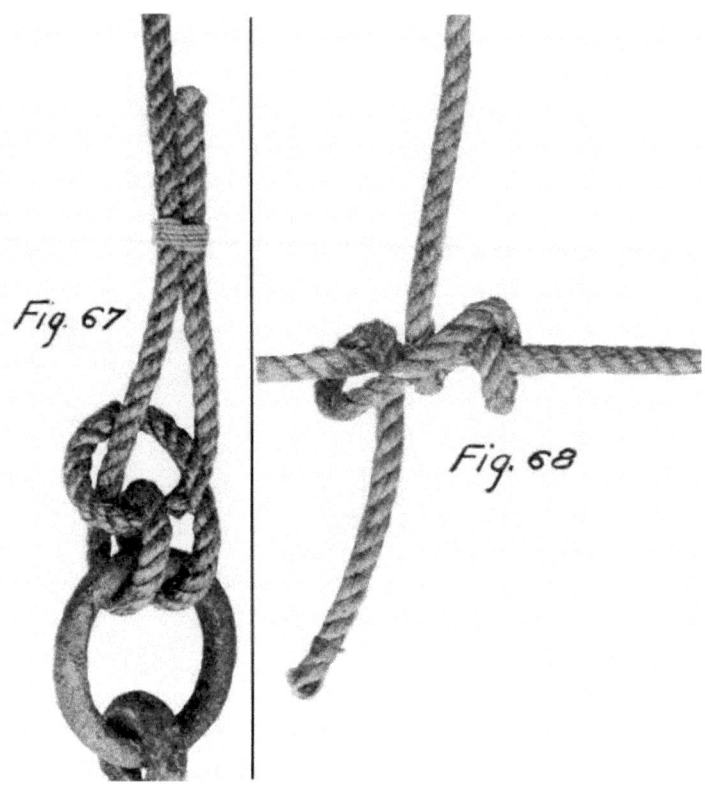

Fig. 67. The Fisherman's-bend is similar to the Fisherman's-hitch except that the half hitches are replaced with whipping.

Fig. 68. A Taut-line-hitch is used for attaching a rope to another rope already under strain, where no slack is available for making any other hitch. It is not secure unless pulled very tight. A few threads of hemp or marlin served about the taut line for the knot to pull against will improve the hitch.

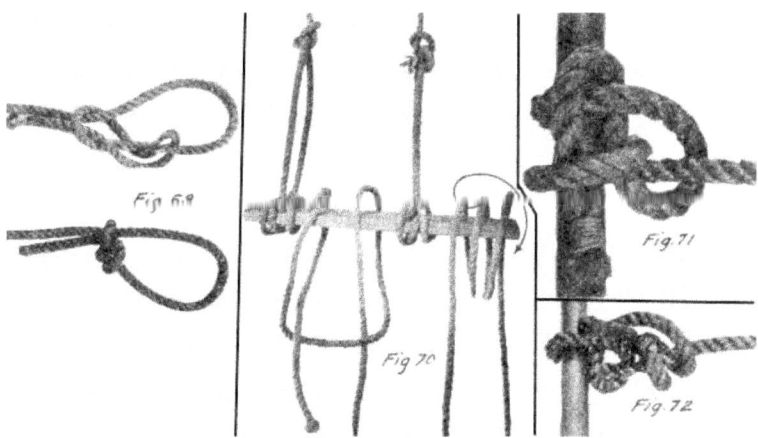

Fig. 69. The Jam Hitch is used in tying up light packages, such as bundles of lath, small boxes, rolls of paper, and the like. It is a hitch that will slide along a cord in one direction, but will jam and hold against moving the other way and will be found exceedingly convenient. The Jam Hitch will answer the requirements provided the cord is large enough and of not too hard a body nor too smooth a surface.

Fig. 70. The Scaffold-hitch is very useful for slinging a scaffold so that it will not turn in the sling. It is started by making a Clove hitch with the two free ends of the rope below the scaffold. Then draw each rope back on itself and up over opposite sides of the board, where the short end is joined to the other with a bowline. Plenty of slack in the Clove will make it possible to draw the bight of each end out to the edge of the scaffold as shown in the left of the figure. The two illustrations at the right of the figure show another method of making a Scaffold hitch. Wrap the rope around the scaffold plank so that it crosses the top of the plank three times. Pull the middle loop as shown by the arrow and fold it down over the end of the plank, resulting as shown in the illustration immediately to the left of the arrow. This is completed by attaching the free end to the standing part with a bowline. Both hitches are equally good.

Fig. 71. The Studding-sail-bend is frequently employed on shipboard for attaching a rope or line to a spar.

Fig. 72. The Midshipman's-hitch is somewhat similar to the Snubbing hitch shown in Figure 54, but is perhaps a little easier to make if the rope is under a strain while being tied.

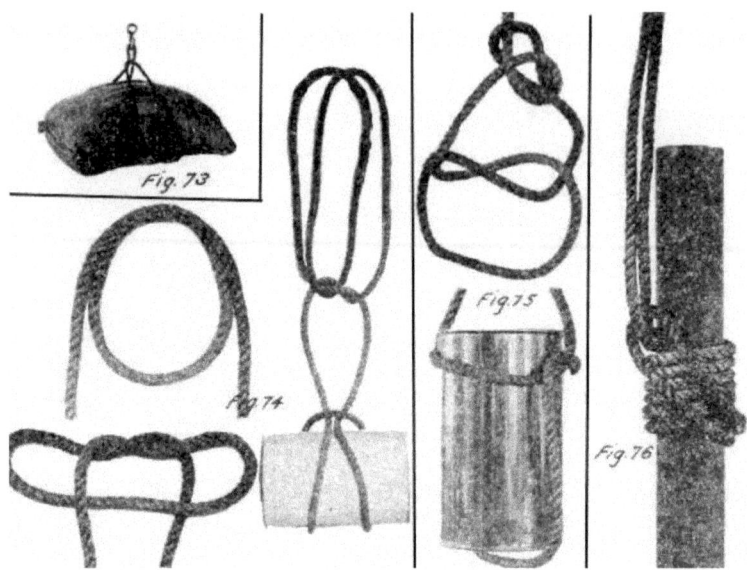

Fig. 73. A Bale-sling as shown is useful where it is necessary to hoist an object to which it is difficult to attach the hoisting tackle. It may be used on bales, sacks, kegs, etc.

Fig. 74. The Hamburger hitch is useful in connection with a bale sling which is too long for the object it is carrying. It is also used to balance the load where two slings are used. The sling is placed around the load as in Fig. 73. Then with the loop end of the sling form a second loop as shown. Where the two ropes cross start to tie a square or Reef knot. Draw up the loops as shown, resulting in the Hamburger hitch. This may be adjusted by running the knot up or down the rope while slack, but it will not slip under strain.

Fig. 75. Sling for a cask, head up, is very useful where it is desired to hoist an open barrel of water or lime or other material. Tie an ordinary knot over the barrel lengthwise. Then separate the two ropes in the middle of the twisted part and drop them over the head of the cask or barrel. Fasten the two rope ends together above the barrel as shown with a bowline.

Fig. 76. A Well Pipe Hitch is used in hoisting pipe, where no special clamp is available for attaching the hoisting tackle to the pipe. The hitch shown will pull tighter, the harder the strain, and is also easy to untie. Pull up all slack possible in the coils when forming the hitch, in order to prevent slipping when the strain is first applied.

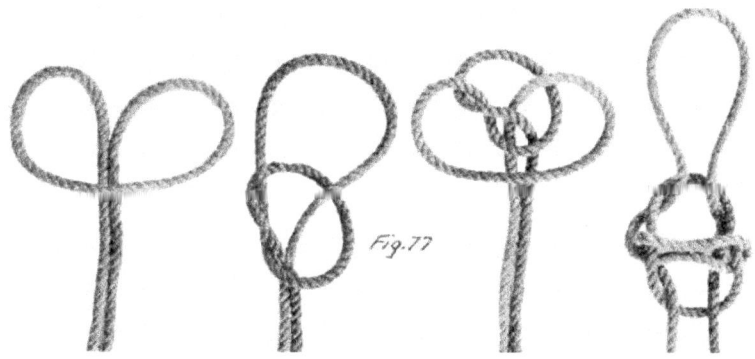

Fig. 77. The Hackamore hitch is commonly known and used as an emergency rope bridle or halter, in the western part of the United States. Among sailors it is known as a running turk's head, and it may be used in carrying a jug or other vessel of similar shape. When used for a halter about twenty feet will be required. The knot is started by forming a bight in the center of the rope. Proceed as indicated in the successive illustrations shown. The result will be a running turk's head. Draw together the two center ropes forming a bridle complete with bit, nose piece, head piece and reins. Such a bridle is not suitable for continuous use, to be sure, but it will be found useful in an emergency.

Fig. 78. The Halter Tie is a knot preferred by some persons for use in hitching or in tying the halter rope in the stall. If properly set, it is secure and may be used in some cases in place of the underhand bowline knot. The halter tie should never be used around a horse's neck, because if the tie is not set up correctly it forms a slip knot and its use might result in strangulation of the animal. In completing the tie draw the end through and set the knot by pulling first on the short end. This is important. If the long rope is pulled first and the kinks in it are straightened out, the tie forms a slip knot, being simply two half hitches around the rope.

Fig. 79. Horse-hitch or tie is commonly used by farmers and stockmen to tie a horse or cow with a rope, so it will not choke itself. Tie an overhand knot in the standing part of the rope and leave open. Tie another overhand knot or a Stevedore knot in the end of the rope. Loop the rope around the animal's neck and insert the knotted end through the open Overhand knot. This hitch will not slip and choke the animal.

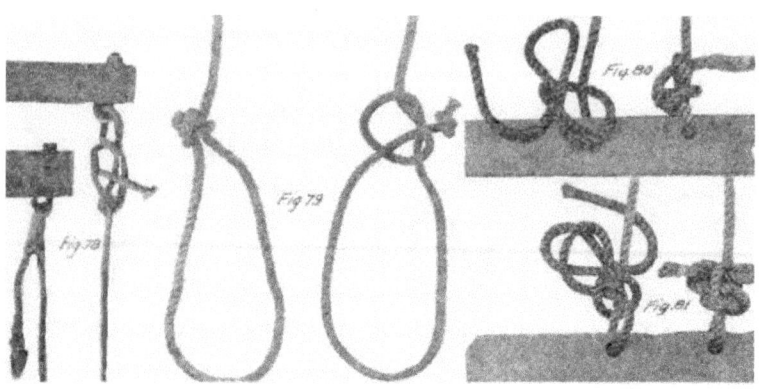

Fig. 80. The Manger tie is used for tying a horse or other animal to a manger or stanchion or hitching rack. The end of the halter rope is first passed through the hole in the manger with a bight or loop on the free end of the rope, tie a slip knot on the standing part. Stick the free end of the rope through the loop or bow as shown. This knot is easily and quickly tied, but under great strain will pull tight, making it hard to untie.

Fig. 81. The Figure Eight Manger Tie is superior to the ordinary Manger Tie in that it will not pull tight under heavy strain such as would occur if the animal became frightened and attempted to break away. Pass the free end of the rope through the hole in the manger or around the hitching rack. Form a bight or loop with the free end of the rope and hold the loop along the standing part. With the free end form another loop and serve around both the first loop and the standing part. Complete the tie by inserting the second loop through the first loop and secure by inserting the free end of the rope through the second loop as shown. This is easily untied by first withdrawing the free end from the loop and then pulling on same until knot is untied.

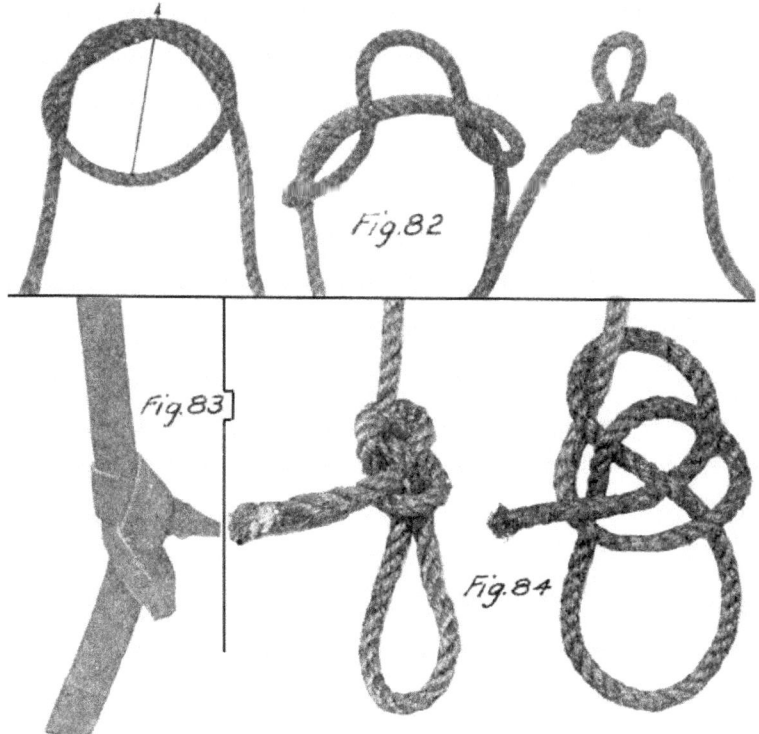

Fig. 82. The Harness hitch is employed for forming a loop on a rope in such a way that strain may be applied to both ends and to the loop without slipping. Start to tie an Over-hand knot as shown. Reach through between the two twisted parts and draw the opposite side of the loop through, following the arrow. The completed harness hitch appears as shown.

Fig. 83. The Strap hitch or Line knot is used to join the free ends of two leather driving lines on a team. It may be employed as an emergency tie for a broken line or strap but is not to be recommended as a permanent repair.

Fig. 84. The Clevis hitch is used for forming a loop on the end of a rope which is both secure under strain and easily untied.

### The Diamond Hitch

The present age of high speed transportation both on land and water, and in the air as well, has served to crowd pack animal transportation back into the hills and into those few regions where rail and sail have not yet penetrated. As a consequence, pack trains are fast becoming unknown, and the skill of the packer is fast being forgotten. The skill of the experienced packer is little short of marvelous, where he can catch a more or less wild pack animal and attach from 100 to 400 or 500 pounds of stuff to his back so securely that it will ride all day without coming off. Different types of freighting, of course, gave rise to different methods of binding on the load, but the more widely used was, no doubt, some form of the famous Diamond Hitch. The early trappers of the Hudson's Bay Company are credited with introducing the Diamond Hitch among the North-West Indians, and old trappers of the period of 1849, engaged in freighting to California, claim that the Mexicans used it at that time.

Different packers have modified and used the Diamond Hitch to suit their needs. As an example, in rough country where there is frequent trouble with pack animals falling with their load, some packers tie the Diamond Hitch so that the final knot is on top of the animal's back where it can be easily reached and loosened with the animal down. Under more favorable conditions, other packers use a Diamond Hitch in which the final tie is made on the side of the animal near the cinch hook. In fact, out of a group of old packers from different localities, the probability is that no two would tie the Diamond Hitch alike in every particular.

The following illustrations of the Diamond Hitch are shown only as types actually in use by different men in the packing business. Other packers may have different methods of tying it more suited to the type of load they are handling. The cuts shown represent the appearance of the Diamond Hitch if the cinch were cut under the animal's belly and the pack were flattened out and laid on the floor with the ropes undisturbed. This method clearly shows in one picture all the different parts of the hitch, so that those interested may follow it in making the hitch for themselves. The Government uses a Spanish packsaddle, or what is known as an aparejo—pronounced, ap-pa-ray-ho, but civilian packers often use the cross tree saddle. It consists of a padded board resting on each side of the animal's backbone. These two padded boards are usually fastened together with two cross trees resembling a saw buck. There are different methods of placing the load on the saddle preparatory to lashing it fast with the Diamond Hitch. No attempt will be made to give complete instructions in packing. The hitches shown are given with the hope they will serve the prospective camper on his vacation to a retreat in the hills, or perhaps satisfy the interest of those who have heard of the Diamond Hitch but have never seen it tied.

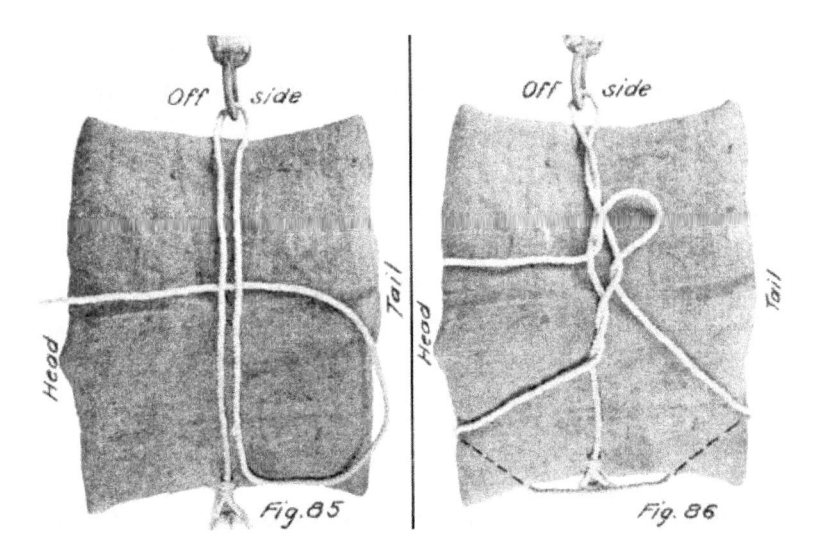

### The Two Man Diamond Hitch

Fig. 85. The Two Man Diamond Hitch is started by laying the middle of the rope lengthwise over the pack from head to tail with the free end of the rope at the head of the animal. Then the cinch hook is thrown under the animal's belly and caught by the off packer. The near packer throws a bight over the pack and the off packer catches it in the cinch hook. The near packer pulls up on the rope, making it tight over the pack.

Fig. 86. The two ropes over the pack are then twisted one and a half times and a loop pulled through as shown. In this case the loop first formed between the rope lying lengthwise and the part crossing the pack is lowered over the near side of the pack.

Fig. 87.

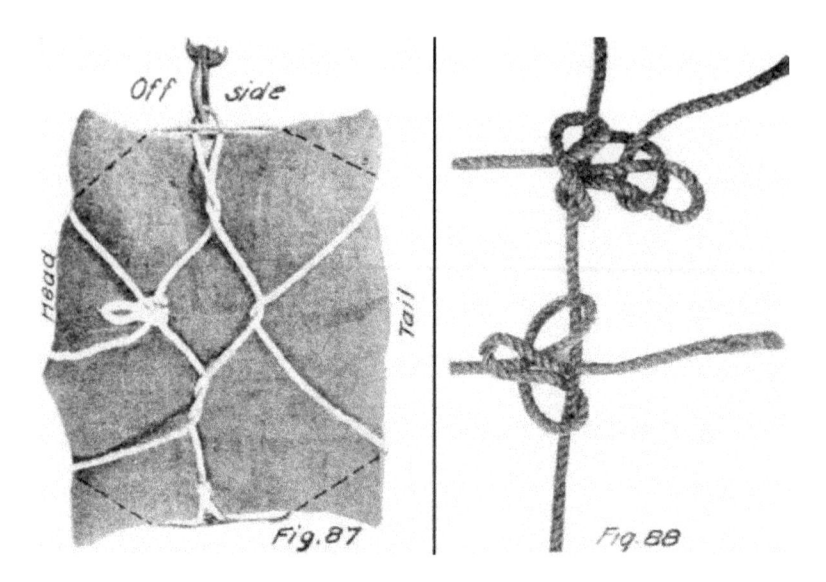

Fig. 88.

Fig. 87. The hitch is then completed by the off packer, as shown. The difference between the one-man hitch and the two-man hitch is that they finish up on different sides of the animal. In the two Diamond Hitches shown, the final tightening pull is taken toward the head of the animal. Many packers tie the Diamond Hitch so that the final pull is taken to rearward of the animal. This can be done by laying the middle of the rope lengthwise of the pack with the end to the rear instead of toward the front of the animal.

Fig. 88. The packer's knot as shown consists of a clove hitch made around a standing rope. The second half hitch is made with a bight instead of the end of the rope. One or more half hitches are then thrown over this loop to make it secure. This knot, if pulled tight in making, will hold very securely, without slipping, and is easily untied by loosening the half hitches, and pulling on the free end of the rope.

### The One Man Diamond Hitch

Fig. 89. The one man Diamond Hitch is employed by one packer working alone and requires that he make two trips around the animal in tying it. The rope is braided into a ring on one end of the cinch. The other end of the cinch carries a hook. Standing on the near side of the animal at its shoulder he first lays the middle of the rope across the pack from forward to back with the free end of the rope forward. He then throws the cinch over the pack and catches the hook under the animal's belly. A loop of the rope is then caught under the cinch hook and pulled tight. Some packers, in using the one man Diamond Hitch, find it helps to hold the hitch tight if they take a double turn around the hook in making the first tightening.

- 34 -

Fig. 89

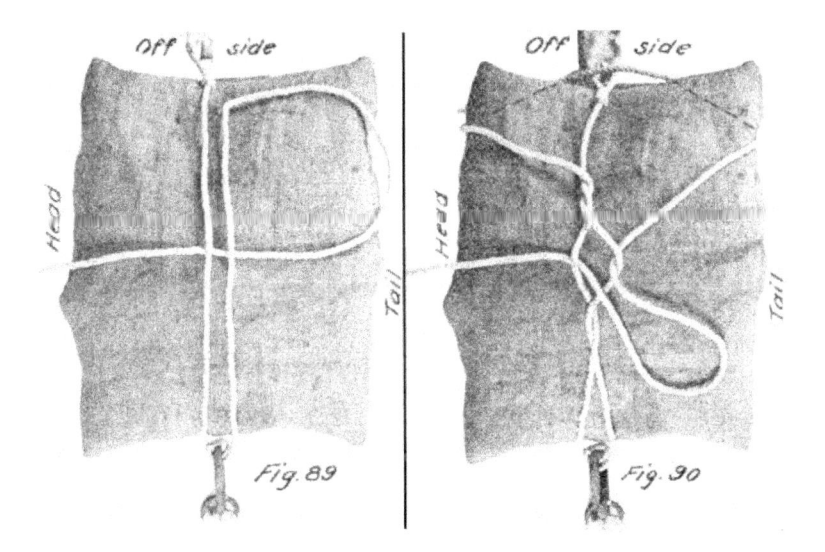

Fig. 90

Fig. 90. Proceeding with the hitch, the two ropes over the pack crosswise are then twisted, lifting the forward strand up and back and pulling the rear strand forward and under. Two turns are made and then a loop of the rope lying forward and back over the top of the pack is drawn up between the two twisted ropes as shown. The loop formed on the off side between the part crosswise of the pack and the part of the rope crossing lengthwise of the pack, is formed over both corners of the off side of the pack. Then the loop drawn up between the two twisted ropes is lowered over the corners of the near side of the pack.

Fig. 91. The final strain is taken on the free end of the rope passing along the neck of the animal and tied at the forward point of the diamond with a packer's knot. If the animal should fall on either side, the knot is easily reached and untied. The free end of the rope is tucked under some part of the hitch or looped over the pack or otherwise disposed of. In making the Diamond Hitch, at no time is the end of the rope pulled through anywhere. This makes it easy to take off without becoming snarled.

Fig. 91

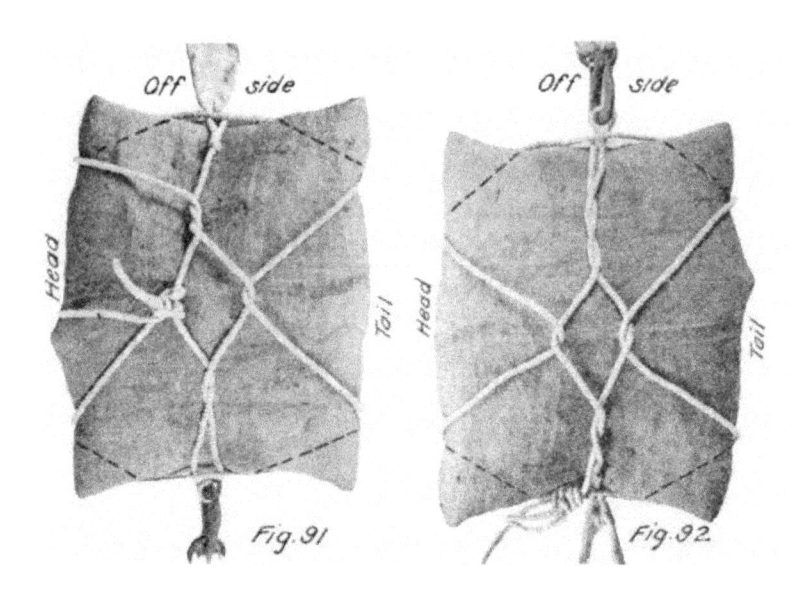

Fig. 92

Fig. 92. The Diamond Hitch as mentioned above is frequently tied so that the knot occurs on the side of the animal opposite the cinch hook instead of on top. This hitch is tied so that the first loop is lowered over the rear corner only of the off side of the pack. In the two other hitches described above, the first loop included both corners of the pack, and finished with a knot on top. The Diamond Hitch shown is thrown by two packers.

# LASHINGS

Fig. 93. To lash a Transom to an upright Spar with the transom in front of the upright. A clove hitch is made around the upright a few inches below the transom. The lashing is brought under the transom, up in front of it, horizontally behind the upright, down in front of the transom, and back behind the upright at the level of the bottom of the transom and above the clove hitch. The following turns are kept outside the previous ones on one spar and inside on the other, not riding over the turns already made. Four turns or more are required. A couple of frapping turns are then taken between the spar and transom, around the lashing, and the lashing is finished off either around one of the spars or any part of the lashing through which the rope can be passed. The final clove hitch should never be made around the spar on the side toward which the stress is to come, as it may jam and be difficult to remove. The lashing must be well beaten with handspike or pick handle to tighten it up. This is called a square lashing.

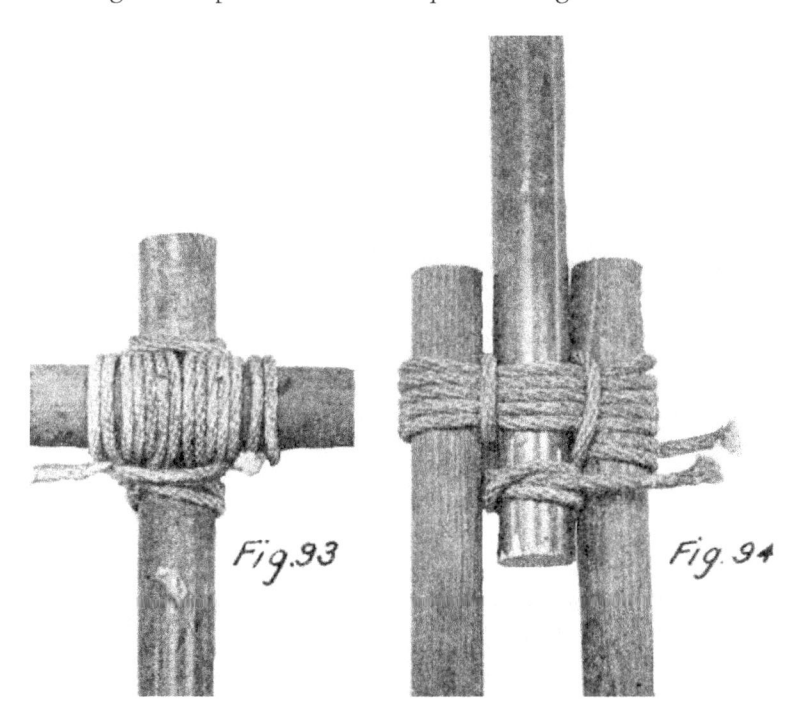

Fig. 93.     Fig. 94.

Fig. 94. To lash three spars together as for a Gin or Tripod. Mark on each spar the distance from the butt to the center of the lashing. Lay two of the spars parallel to each other with an interval a little greater than the diameter. Rest their tips on a skid and lay the third spar between them with its butt in the opposite direction so that the marks on the three spars will be in line.

Make a clove hitch on one of the outer spars below the lashing and take eight or nine loose turns around the three, as shown in Figure 94. Take a couple of frapping turns between each pair of spars in succession and finish with a clove hitch on the central spar above the lashing. Pass a sling over the lashing and the tripod is ready for raising.

# TACKLE SETS

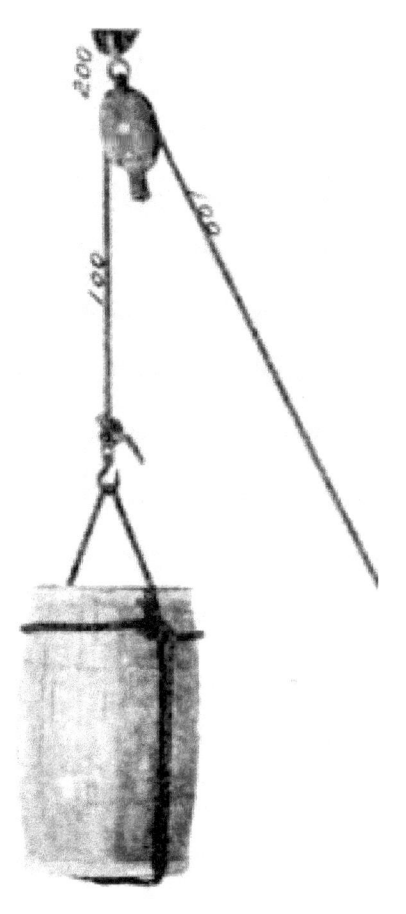

Fig. 95.

The use of block and tackle affords at least two advantages to the user. One is the advantage of position. The user may stand on the ground and pull downward—the most easy and natural way of exerting force, while the resulting forces may be developed upward as in the case of a hoist. The other advantage is mechanical. By the use of a combination of lines and sheaves, force applied by the user can be multiplied many times before it is transferred to act upon the body. But where there is gain in pounds force applied, there is always a counteracting loss due to an increase in the distance required to apply the force compared with the distance the weight or load will travel; as in Figure 96, a force of 100 lbs. on the free end of the rope will give a resultant on the object of 200 lbs. (neglecting loss by friction in rope and pulley) but distance travelled by the user will be two feet to one foot travelled by the object.

The illustrations are shown in each case with an arbitrary force of 100 lbs. applied to the free end of the rope. The resulting force (neglecting or disregarding friction) is then shown in all parts of the set. In actual practice the friction of the sheave and the resistance of the rope to bending gives rise to a loss of about 5% of the force applied to the rope passing through each sheave. For example in Fig. 95 the force applied on the barrel would be 95% of that applied to the free end of the rope or 95 lbs. In Fig. 96 the resultant force would be 100 + (100 - 5) = 195 lbs. and in Fig. 97, the lift on the armature would be 185½ lbs. instead of 200 as shown.

The ropes are also separated in the illustrations in order to show each part clearly. The ropes are assumed to pull parallel to each other and the figures represent the pounds resulting in different parts of the set under those conditions. The illustrations show some of the most typical applications of block and tackle for mechanical advantage or advantage of position or both.

Fig. 95. The Single Whip affords only advantage of position commonly used on a crane or derrick or perhaps for hauling an object up to a wall or to the water's edge. Theoretical advantage 1:1.

Fig. 96.

Fig. 96. The Running tackle is similar to the Single Whip except that the object to be moved is attached at a different place. This gives a theoretical advantage of 2:1.

Fig. 97. The Gun tackle A affords an advantage of position since the user stands on the ground and pulls down and the resultant force is applied to the object vertically upward. Theoretical advantage 2:1.

Fig. 98. The gun tackle B is the same as gun tackle A except that its application is different, giving a theoretical advantage of 3:1.

Fig. 99. Whip-on-whip multiplies the mechanical advantage by two, where applied as shown. If inverted and the top block applied to the load with the loop snubbed the mechanical advantage would be 4:1. In both cases two single blocks are used.

Fig. 100. The Luff tackle has many applications aside from the one shown. Ordinarily consisting of one single and one double block and a single rope, it gives a theoretical mechanical advantage of 4:1 in the case shown.

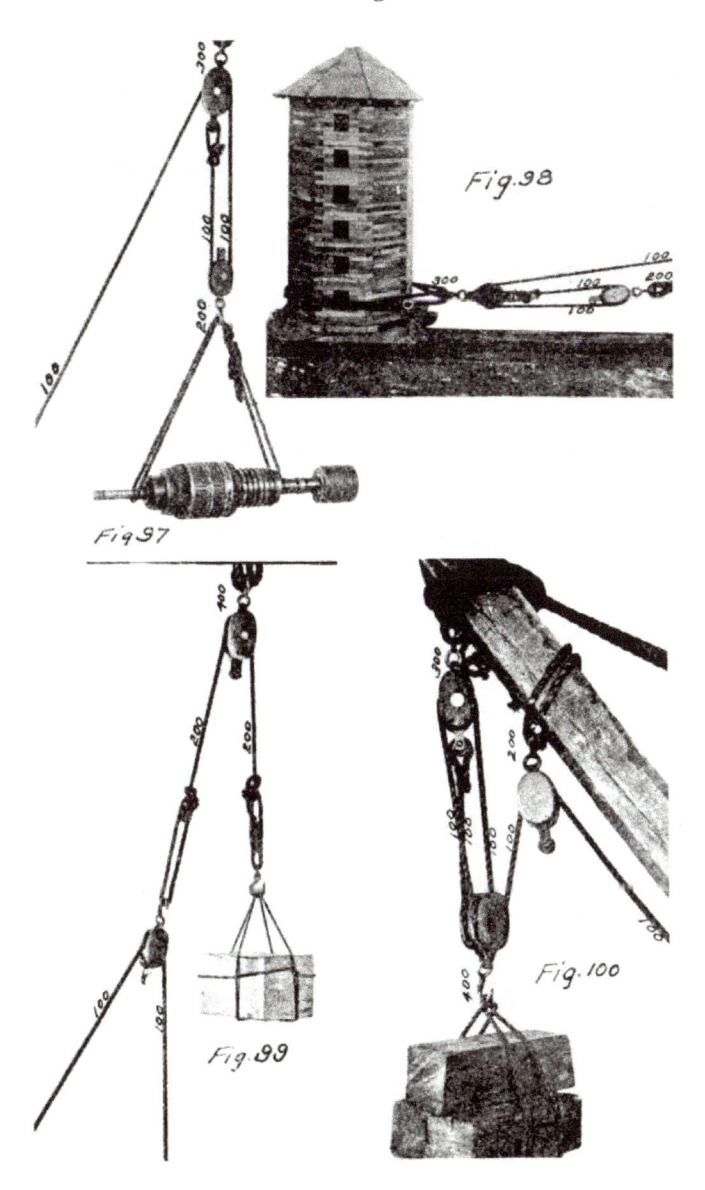

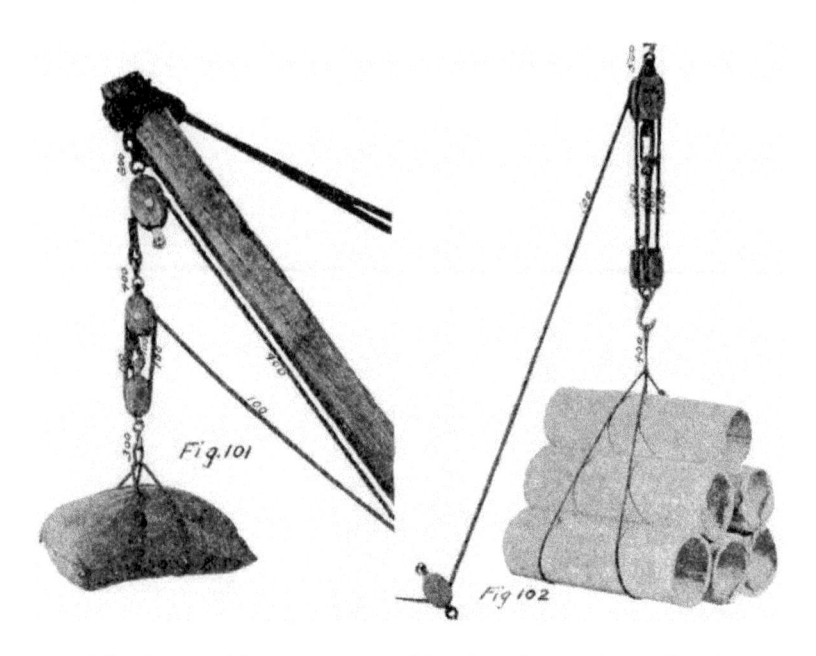

Fig. 101. The Port tackle, consisting of Single Whip and a Luff tackle may be applied when the level of operations changes from time to time and it is undesirable to apply the amount of rope necessary to make the Luff part of the set long enough to serve for all levels. A bale sling is also shown in use.

Fig. 102. A Double Luff tackle has a four part line instead of a three part line as in the Single Luff.

Fig. 103. A Single Spanish Burton (A) using two single blocks and one rope gives a greater mechanical advantage than the same equipment used as in Figure 97, the Gun Tackle. This is useful in shifting cargo, etc., where the distance hoisted in not great.

Fig. 104. A Single Spanish Burton (B) using three single blocks and two ropes, gives the same hoisting range as the Type A Burton, but a greater mechanical advantage.

Fig. 105. Three Fold Purchase using a six part line, gives a theoretical mechanical advantage of 6:1 and an actual advantage of 5·03:1, assuming a loss of 5% of the force on the rope passing over each sheave.

Fig.103

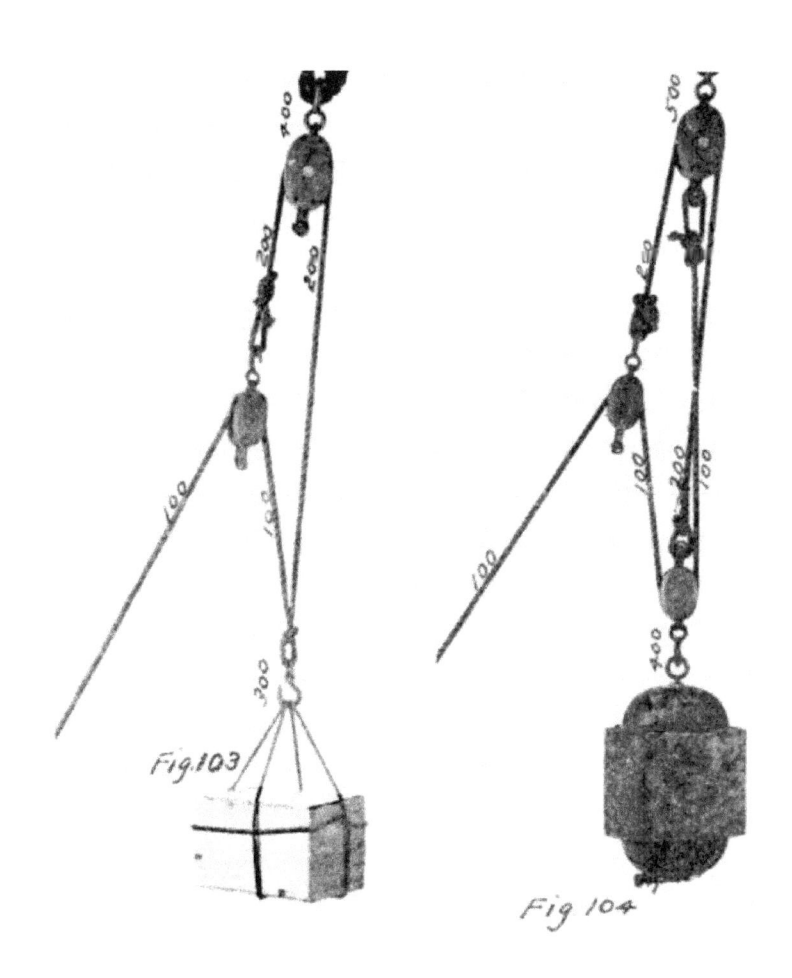

Fig 104

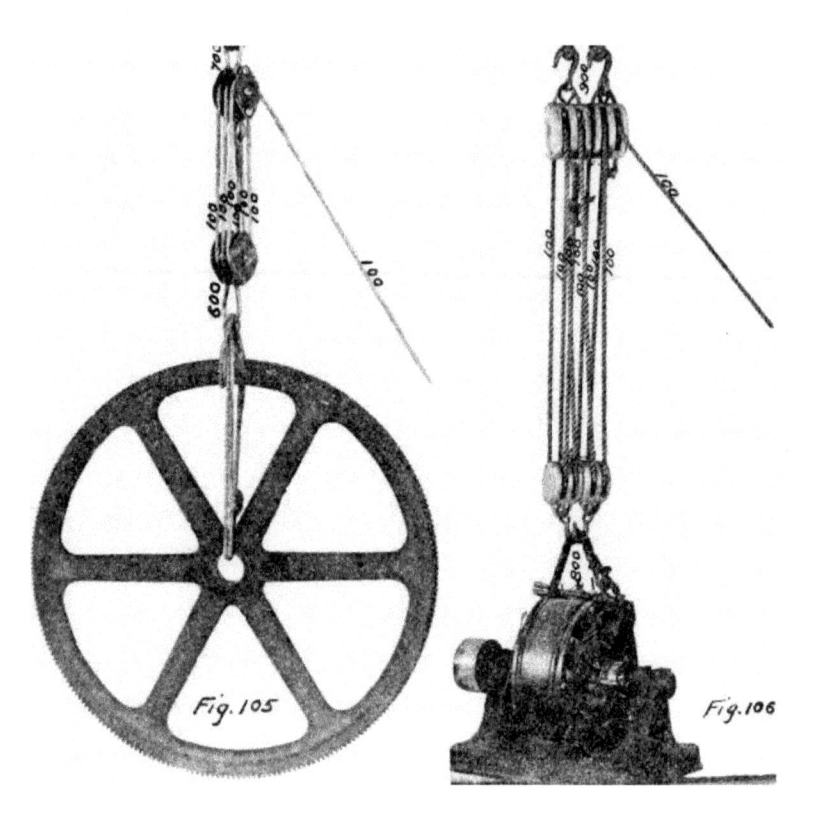

Fig. 106. Four Fold purchase using two four-sheave blocks, is commonly used in derricks and hoists. The illustration shows the possibility of using four two-sheave blocks, where the larger sizes are not available.

Fig. 107. The Double Burton (A), for one rope and two single blocks and one double block, gives a limited hoisting range which is desirable in shifting heavy weights when it is necessary to lift them but a small distance.

Fig. 108. The Double Burton (B), while using exactly the same equipment as is used in Fig. 91, shows the large differences in mechanical advantage between different methods of threading up the set. The illustration also shows a box sling in use.

Fig. 109. Double Burton (C), is a further application of the principle of the Spanish Burton, using two ropes.

Fig. 107

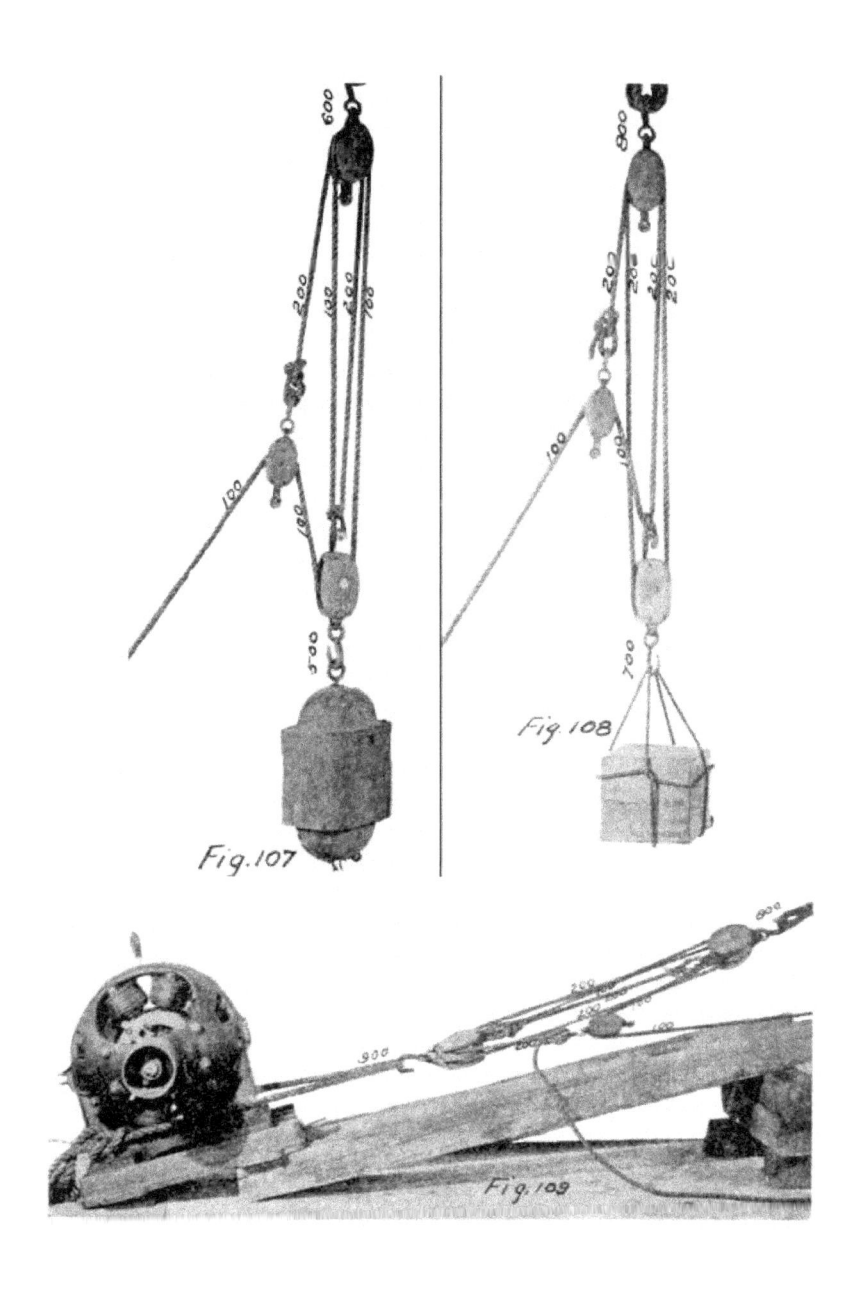

Fig. 108

Fig. 109

Fig. 110.

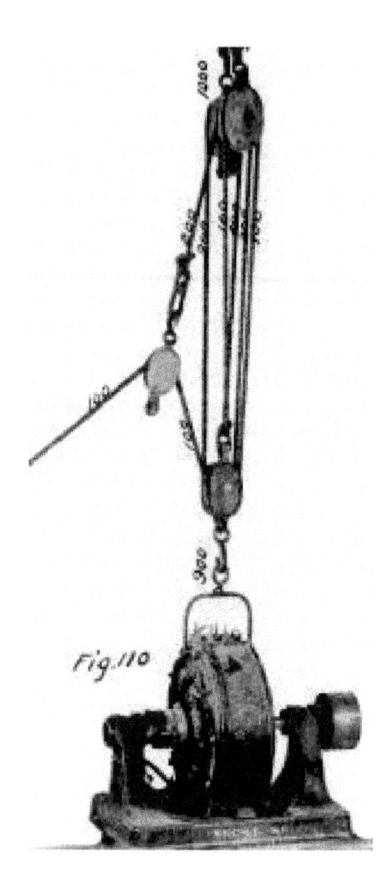

Fig. 110. Double Burton (D), using but one rope, illustrates the possiblity of using it to greater mechanical advantage than would be possible in a six fold purchase. However, in this case the hoisting range is less than would be possible in a six fold purchase.

Fig. 111. Luff on Luff illustrates a common application of tackle to secure mechanical advantage. It will readily be recognized that the major tackle must be four times as strong as the other set if both are to be used anywhere near to capacity.

Fig. 112. Another Double Burton which also illustrates the possibility of combining two blocks in place of one, with the required number of sheaves.

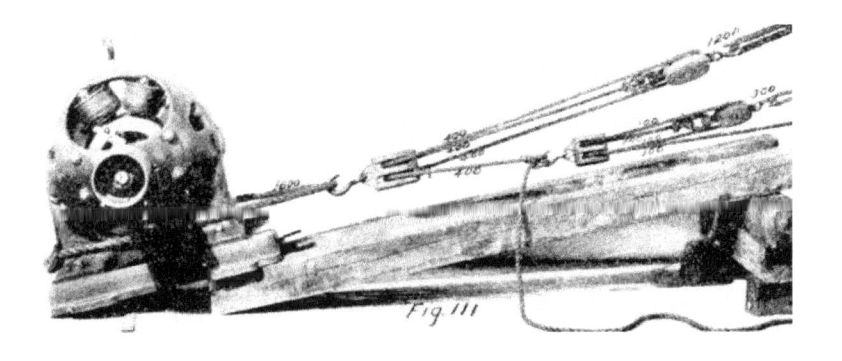

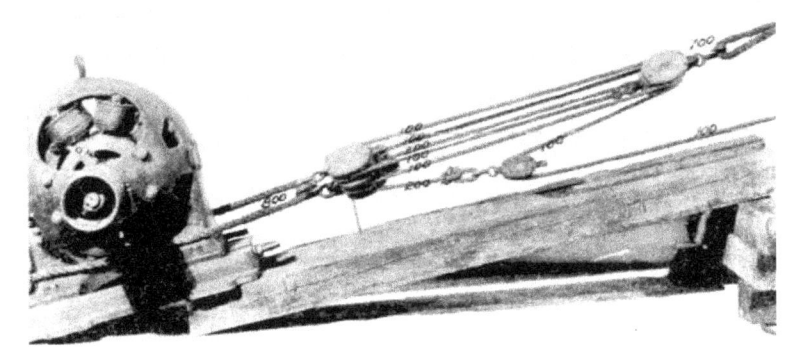

Fig. 112.

# CHAIN HOISTS

Frequent use is made in garages, machine shops and other places, of a special device for hoisting heavy machine parts. The apparatus referred to is known as a chain hoist. These are built to use chain instead of rope and are designed to operate slowly, but with great mechanical advantage. Different types embody different design of movements, some being merely a train of gears attached to a sheave wheel and driven by a worm gear. Others employ the differential principle in which the hoisting chain is double, one end running over a small pulley and the other end running in the opposite direction over a larger pulley on the same shaft. As the small pulley unwinds one end of the chain slowly, the other pulley winds up the other end faster—thus raising the lower end of the chain loop. Chain hoists are made for various capacities, and can be built to raise the load any desired distance, merely by supplying chain long enough. A chain-hoist attached to a travelling crane makes a very satisfactory equipment for a shop where heavy parts are to be lifted and transferred and should be used wherever there is enough such work to warrant the greater first cost.

Fig. 113. A Geared-chain-hoist showing a 1-ton hoist manufactured by the Wright Mfg. Co., of Lisbon, Ohio, using two chains, one for lifting and the other for operating.

Fig. 114. A Differential Chain hoist using a single continuous chain running through a pulley at the bottom and over two different sized wheels fastened on the same shaft at the top. As one unwinds the other winds up and the difference in diameter causes one to wind up faster than the other unwinds.

Fig. 115. A Chinese hoist or Chinese capstan, in which the differential principle is used. The illustration shows the possibility of quickly applying the principle to the hoisting of a well-casing. It has the merit of being cheap and easy to construct and very efficient in developing a large mechanical

- 49 -

advantage. The necessary materials can frequently be found around almost any farm or construction camp.

Fig. 116. A Snatch Block is used frequently in connection with hay handling equipment on the farm. Hoisting hay from a loaded wagon to the track located in the peak of the barn, requires much more force than is required to move the load along the track. From then on, the snatch block pulls away from the knot causing the load to travel on the carrier track twice as fast as the team. The object is to utilize the direct pull of the team while elevating the load and increase the speed of the load and decrease the distance travelled by the team after the load has been elevated and is to be transferred.

# TRANSMISSION CABLES

### Hemp and Manila

Ropes and cables have many uses and applications both in industry and pleasure. Haulage, hoisting and the transmission of power are three of the most modern applications to which ropes and cables have been put, which require an intimate knowledge of their strength and life in service, in order to secure satisfactory service. For instance, a certain kind and size of rope is suitable for guy lines but would not be able to compete with a different type of rope in service on a rapid hoist. Similarly, a certain size of rope is being used on a rope drive, but the power load is increasing to such a point it is necessary to increase the size of transmission rope. If the sheaves are not increased in diameter suitable to the increased size of rope, the acute bending of the larger rope on the old sheave wheel will shorten its life materially.

Following are tables of strength for a few different kinds and sizes of ropes. It is not the purpose to make these tables complete and exhaustive in scope, but rather to give a general conception of the strength to be expected of different kinds and sizes of ropes in more common use. Those interested in more complete information on this subject should refer to the catalogs put out by manufacturers of ropes.

No accurate rule can be given for calculating the strength of rope and any table giving the strength will only be approximately correct. Four-strand rope has about 16% more strength than three-strand rope. Tarring rope decreases the strength by about 25% because the high temperature of the tar injures the fibers. The strength of a rope is decreased by age, exposure and wear.

The breaking strength of a rope is the weight or pull that will break it. The safe load is the weight you may put on a rope without danger of breaking it. The safe load must be very much less than the breaking strength, in order that life and property may not be endangered when heavy objects are to be moved or lifted. The safe load is usually regarded as 1/6 of the breaking strength. The breaking strength and safe load for all ropes must be largely a matter of good judgment and experience.

### Calculation of Strength

For new manila rope the breaking strength in pounds may be found approximately by the following rule: Square the diameter, measured in inches, and multiply this product by 7200. Result obtained from this rule may vary as much as 15% from actual tests. The safe load can be found by dividing the breaking strength by 6.

Hemp rope is approximately 3/4 as strong as manila so that we use the following rule for it: The breaking strength of hemp rope in pounds is 5400 times the square of the diameter in inches. The safe load is found by dividing the breaking strength by 6 as we did for the manila rope.

## Care of Rope

Keep rope in a dry place, do not leave it out in the rain. If a rope gets wet, stretch it out straight to dry. Do not let the ends become untwisted but fix them in some way to prevent it as soon as the rope is obtained. A stiff and hard rope may be made very soft and flexible by boiling for a time in pure water. This will of course remove some of the tar or other preservative. Cowboys treat their lasso ropes in this way.

## Uncoiling Rope

1. Start with the end found in the center of the coil.

2. Pull this end out and the rope should uncoil in a direction opposite to the direction of motion of the hands of a clock.

3. If it uncoils in the wrong direction, turn the coil over and pull this same end through the center of the coil and out on the other side.

4. If these directions are followed, the rope will come out of the coil with very few kinks or snarls.

## SIZE AND STRENGTH OF TEXTILE ROPES

| Diam. of Rope Inches | Ultimate Strength, Lb. ||| Working Strength, Lbs. |||
|---|---|---|---|---|---|---|
| | Cotton | Manila | Hemp | Cotton | Manila | Hemp |
| 1/2 | 1,150 | 1,900 | | 50 | 50 | |
| 5/8 | 1,800 | 2,900 | | 78 | 78 | |
| 3/4 | 2,600 | 4,100 | | 112 | 112 | |
| 7/8 | 3,500 | 5,500 | | 153 | 153 | |
| 1 | 4,600 | 7,100 | | 200 | 200 | |
| 1 1/4 | 7,200 | 10,900 | | 312 | 312 | |
| 1 1/2 | 10,400 | 15,000 | | 450 | 450 | |
| 1 3/4 | 14,000 | 19,800 | | 612 | 612 | |
| 2 | 18,400 | 25,100 | | 800 | 800 | |

## STRENGTH OF MANILA ROPE

| Diam. of Rope in Inches | Average Quality New Manila Rope |
|---|---|
| 2 3/4 | 26 |
| 2 1/2 | 21 1/2 |
| 2 1/4 | 18 1/2 |
| 2 | 15 |
| 1 3/4 | 12 1/2 |
| 1 5/8 | 10 |
| 1 1/2 | 8 1/2 |
| 1 3/8 | 7 1/2 |
| 1 1/4 | 6 1/4 |
| 1 1/8 | 5 1/4 |
| 1 | 4 |
| 7/8 | 3 1/4 |
| 3/4 | 2 1/4 |
| 5/8 | 2 |
| 9/16 | 1 1/2 |
| 1/2 | 1 1/5 |
| 7/16 | 3/4 |
| 3/8 | 1/2 |
| 5/16 | 3/8 |
| 9/32 | 3/10 |
| 1/4 | 1/4 |

# STEEL CABLES

The modern demands of industry for speed and large capacity have called for strengths exceeding that possible to attain from hemp or manila ropes, which are not excessive in size or cost. As a result, steel ropes and cables have been developed and perfected to a high degree of strength and dependability. The majority of hoists and cranes use steel rope. Logging industries depend for most part on steel cables. Cable cars use special steel cables which in many cases are several miles long. Long tramways use light steel cables, for long spans where manila rope would scarcely maintain its own weight. High speed passenger elevators maintain safe and dependable service day after day only through the strength of the perfected flexible steel cable. However, as stated above, each particular type of service calls for some special type of cable. The following tables are not complete but will serve to indicate the scope of the field covered by this subject.

## CAST STEEL ROPE

Composed of 6 strands and a hemp center, 7 wires to the strand

| Diameter in Inches | Approximate Circumference in Inches | Approx. Breaking Strain in Tons of 2000 lbs. | Proper Working Load in Tons of 2000 lbs. | Minimum Size of Drum or Sheave in ft. |
|---|---|---|---|---|
| 1 1/2 | 4 3/4 | 63 | 12.6 | 11 |
| 1 3/8 | 4 1/4 | 53 | 10.6 | 10 |
| 1 1/4 | 4 | 46 | 9.2 | 9 |
| 1 1/8 | 3 1/2 | 37 | 7.4 | 8 |
| 1 | 3 | 31 | 6.2 | 7 |
| 7/8 | 2 3/4 | 24 | 4.8 | 6 |
| 3/4 | 2 1/4 | 18.6 | 3.7 | 5 |
| 11/16 | 2 1/8 | 15.4 | 3.1 | 4 3/4 |
| 5/8 | 2 | 13 | 2.6 | 4 1/2 |
| 9/16 | 1 3/4 | 10 | 2 | 4 |
| 1/2 | 1 1/2 | 7.7 | 1.54 | 3 1/2 |

| Diameter in Inches | Approximate Circumference in Inches | Approx. Breaking Strain in Tons of 2000 lbs. | Proper Working Load in Tons of 2000 lbs. | Minimum Size of Drum or Sheave in ft. |
|---|---|---|---|---|
| 7/16 | 1 1/4 | 5.5 | 1.10 | 3 |
| 3/8 | 1 1/8 | 4.6 | .92 | 2 3/4 |
| 5/16 | 1 | 3.5 | .70 | 2 1/4 |
| 9/32 | 7/8 | 2.5 | .50 | 1 3/4 |

## CAST STEEL ROPE

Composed of 6 strands and a hemp center, 19 wires to the strand

| Diam. of Rope in Inches | Approximate Circumference in Inches | Approx. Breaking Strain in Tons of 2000 lbs. | Proper Working Load in Tons of 2000 lbs. | Minimum Size of Drum or Sheave in ft. |
|---|---|---|---|---|
| 2 3/4 | 8 5/8 | 211 | 42.2 | 11 |
| 2 1/2 | 7 7/8 | 170 | 34 | 10 |
| 2 1/4 | 7 1/8 | 133 | 26.6 | 9 |
| 2 | 6 1/4 | 106 | 21.2 | 8 |
| 1 7/8 | 5 3/4 | 96 | 19 | 8 |
| 1 3/4 | 5 1/2 | 85 | 17 | 7 |
| 1 5/8 | 5 | 72 | 14.4 | 6 1/2 |
| 1 1/2 | 4 3/4 | 64 | 12.8 | 6 |
| 1 3/8 | 4 1/4 | 56 | 11.2 | 5 1/2 |
| 1 1/4 | 4 | 47 | 9.4 | 5 |
| 1 1/8 | 3 1/2 | 38 | 7.6 | 4 1/2 |
| 1 | 3 | 30 | 6 | 4 |
| 7/8 | 2 3/4 | 23 | 4.6 | 3 1/2 |

| Diam. of Rope in Inches | Approximate Circumference in Inches | Approx. Breaking Strain in Tons of 2000 lbs. | Proper Working Load in Tons of 2000 lbs. | Minimum Size of Drum or Sheave in ft. |
|---|---|---|---|---|
| 3/4 | 2 1/4 | 17.5 | 3.5 | 3 |
| 5/8 | 2 | 12.5 | 2.5 | 2 1/2 |
| 9/16 | 1 3/4 | 10 | 2 | 2 1/4 |
| 1/2 | 1 1/2 | 8.4 | 1.68 | 2 |
| 7/16 | 1 1/4 | 6.5 | 1.30 | 1 3/4 |
| 3/8 | 1 1/8 | 4.8 | .96 | 1 1/2 |
| 5/16 | 1 | 3.1 | .62 | 1 1/4 |
| 1/4 | 3/4 | 2.2 | .44 | 1 |

## CAST STEEL ROPE

Composed of 6 strands and a hemp center, 37 wires to the strand

| Diameter in Inches | Approximate Circumference in Inches | Approx. Breaking Strain in Tons of 2000 lbs. | Proper Working Load in Tons of 2000 lbs. | Minimum Size of Drum or Sheave in ft. |
|---|---|---|---|---|
| 2 3/4 | 8 5/8 | 200 | 40 | |
| 2 1/2 | 7 7/8 | 160 | 32 | |
| 2 1/4 | 7 1/8 | 125 | 25 | |
| 2 | 6 1/4 | 105 | 21 | |
| 1 3/4 | 5 1/2 | 84 | 17 | |
| 1 5/8 | 5 | 71 | 14 | |
| 1 1/2 | 4 3/4 | 63 | 12 | 3 3/4 |
| 1 3/8 | 4 1/4 | 55 | 11 | 3 1/2 |

| Diameter in Inches | Approximate Circumference in Inches | Approx. Breaking Strain in Tons of 2000 lbs. | Proper Working Load in Tons of 2000 lbs. | Minimum Size of Drum or Sheave in ft. |
|---|---|---|---|---|
| 1 1/4 | 4 | 45 | 9 | 3 1/4 |
| 1 1/8 | 3 1/2 | 34 | 6.8 | 2 3/4 |
| 1 | 3 | 29 | 5.8 | 2 1/2 |
| 7/8 | 2 3/4 | 23 | 4.6 | 2 1/4 |
| 3/4 | 2 1/4 | 17.5 | 3.5 | 1 3/4 |
| 5/8 | 2 | 11.2 | 2.2 | 1 3/4 |
| 9/16 | 1 3/4 | 9.5 | 1.9 | 1 1/2 |
| 1/2 | 1 1/2 | 7.25 | 1.45 | 1 1/4 |
| 7/16 | 1 1/4 | 5.50 | 1.10 | 1 1/4 |
| 3/8 | 1 1/8 | 4.20 | .84 | 1 |

## CAST STEEL ROPES FOR INCLINES

Six strands of 7 wires each—hemp center

| Diam. of Rope in Inches | Diam. of Sheaves or Drums in Feet, Showing Percentage of Life for Various Diams. ||||||| 
|---|---|---|---|---|---|---|---|
| | 100 % | 90 % | 80 % | 75 % | 60 % | 50 % | 25 % |
| 1 1/2 | 16 | 14 | 12 | 11 | 9 | 7 | 4.75 |
| 1 3/8 | 14 | 12 | 10 | 8.5 | 7 | 6 | 4.5 |
| 1 1/4 | 12 | 10 | 8 | 7.25 | 6 | 5.5 | 4.25 |
| 1 1/8 | 10 | 8.5 | 7.75 | 7 | 6 | 5 | 4 |
| 1 | 8.5 | 7.75 | 6.75 | 6 | 5 | 4.5 | 3.75 |
| 7/8 | 7.75 | 7 | 6.25 | 5.75 | 4.5 | 3.75 | 3.2 |
| 3/4 | 7 | 6.25 | 5.5 | 5 | 4.25 | 3.5 | 2.75 |

| Diam. of Rope in Inches | Diam. of Sheaves or Drums in Feet, Showing Percentage of Life for Various Diams. |||||||
|---|---|---|---|---|---|---|---|
| | 100 % | 90 % | 80 % | 75 % | 60 % | 50 % | 25 % |
| ⅝ | 6 | 5.25 | 4.5 | 4 | 3.25 | 3 | 2.5 |
| ½ | 5 | 4.5 | 4 | 3.5 | 2.75 | 2 | 1.75 |

## CAST STEEL HOISTING ROPES

6 strands of 19 wires each—hemp center

| Diam. of Rope in Inches | Diam. of Sheaves or Drums in Feet, Showing Percentage of Life for Various Diams. |||||||
|---|---|---|---|---|---|---|---|
| | 100 % | 90 % | 80 % | 75 % | 60 % | 50 % | 25 % |
| 1 ½ | 14 | 12 | 10 | 8.5 | 7 | 6 | 4.5 |
| 1 ⅜ | 12 | 10 | 8 | 7 | 6 | 5.25 | 4.25 |
| 1 ¼ | 10 | 8.5 | 7.5 | 6.75 | 5.5 | 5 | 4 |
| 1 ⅛ | 9 | 7.5 | 6.5 | 5.5 | 5 | 4.5 | 3.75 |
| 1 | 8 | 7 | 6 | 5.5 | 4.5 | 4 | 3.50 |
| ⅞ | 7.5 | 6.75 | 5.75 | 5 | 4.25 | 3.5 | 3 |
| ¾ | 5.5 | 4.5 | 4 | 3.75 | 3.25 | 3 | 2.25 |
| ⅝ | 4.5 | 4 | 3.75 | 3.25 | 3 | 2.5 | 2 |
| ½ | 4 | 3 | 3 | 2.75 | 2.25 | 2 | 1.5 |
| ⅜ | 3 | | | 2 | | 1.5 | |

## STANDARD HOISTING ROPE

Six Strands—19 wires to the strand—One hemp core

## Swedes Iron

| Diam. in Inches | Circum. in Inches | Approx. Weight Per Foot | Approx. Strain in Tons of 2000 lbs. | Proper Working Load in Tons of 2000 lbs. | Diam. of Drum or Sheave in ft. Advised |
|---|---|---|---|---|---|
| 2 ¾ | 8 ⅝ | 11.95 | 111 | 22.2 | 17 |
| 2 ½ | 7 ⅞ | 9.85 | 92 | 18.4 | 15 |
| 2 ¼ | 7 ⅛ | 8 | 72 | 14.4 | 14 |
| 2 | 6 ¼ | 6.30 | 55 | 11 | 12 |
| 1 ⅞ | 5 ¾ | 5.55 | 50 | 10 | 12 |
| 1 ¾ | 5 ½ | 4.85 | 44 | 8.8 | 11 |
| 1 ⅝ | 5 | 4.15 | 38 | 7.5 | 10 |
| 1 ½ | 4 ¾ | 3.55 | 33 | 6.5 | 9 |
| 1 ⅜ | 4 ¼ | 3 | 28 | 5.5 | 8.5 |
| 1 ¼ | 4 | 2.45 | 22.8 | 4.56 | 7.5 |
| 1 ⅛ | 3 ½ | 2 | 18.6 | 3.72 | 7 |
| 1 | 3 | 1.58 | 14.5 | 2.90 | 6 |
| ⅞ | 2 ¾ | 1.20 | 11.8 | 2.36 | 5.5 |
| ¾ | 2 ¼ | .89 | 8.5 | 1.70 | 4.5 |
| ⅝ | 2 | .62 | 6 | 1.20 | 4 |
| 9/16 | 1 ¾ | .50 | 4.7 | .94 | 3.5 |
| ½ | 1 ½ | .39 | 3.9 | .78 | 3 |
| 7/16 | 1 ¼ | .30 | 2.9 | .58 | 2.75 |
| ⅜ | 1 ⅛ | .22 | 2.4 | .48 | 2.25 |
| 5/16 | 1 | .15 | 1.5 | .30 | 2 |
| ¼ | ¾ | .10 | 1.1 | .22 | 1.50 |

## Crucible Cast Steel

| Diam. in Inches | Circum. in Inches | Approx. Weight Per Foot | Approx. Strain in Tons of 2000 lbs. | Proper Working Load in Tons of 2000 lbs. | Diam. of Drum or Sheave in ft. Advised |
|---|---|---|---|---|---|
| 2 3/4 | 8 5/8 | 11.95 | 211 | 42.2 | 11 |
| 2 1/2 | 7 7/8 | 9.85 | 170 | 34 | 10 |
| 2 1/4 | 7 1/8 | 8 | 133 | 26.6 | 9 |
| 2 | 6 1/4 | 6.30 | 106 | 21.2 | 8 |
| 1 7/8 | 5 3/4 | 5.55 | 96 | 19 | 8 |
| 1 3/4 | 5 1/2 | 4.85 | 85 | 17 | 7 |
| 1 5/8 | 5 | 4.15 | 72 | 14.4 | 6.5 |
| 1 1/2 | 4 3/4 | 3.55 | 64 | 12.8 | 6 |
| 1 3/8 | 4 1/4 | 3 | 56 | 11.6 | 5.5 |
| 1 1/4 | 4 | 2.45 | 47 | 9.4 | 5 |
| 1 1/8 | 3 1/2 | 2 | 38 | 7.6 | 4.5 |
| 1 | 3 | 1.58 | 30 | 6 | 4 |
| 7/8 | 2 3/4 | 1.20 | 23 | 4.6 | 3.5 |
| 3/4 | 2 1/4 | .89 | 17.5 | 3.5 | 3 |
| 5/8 | 2 | .62 | 12.5 | 2.5 | 2.5 |
| 9/16 | 1 3/4 | .50 | 10 | 2 | 2.25 |
| 1/2 | 1 1/2 | .39 | 8.4 | 1.68 | 2 |
| 7/16 | 1 1/4 | .30 | 6.5 | 1.30 | 1.75 |
| 3/8 | 1 1/8 | .22 | 4.8 | .96 | 1.50 |
| 5/16 | 1 | .15 | 3.1 | .62 | 1.25 |
| 1/4 | 3/4 | .10 | 2.2 | .44 | 1 |

## Plow Steel

| Diam. in Inches | Circum. in Inches | Approx. Weight Per Foot | Approx. Strain in Tons of 2000 lbs. | Proper Working Load in Tons of 2000 lbs. | Diam. of Drum or Sheave in ft. Advised |
|---|---|---|---|---|---|
| 2 3/4 | 8 5/8 | 11.95 | 275 | 55 | 11 |
| 2 1/2 | 7 7/8 | 9.85 | 229 | 46 | 10 |
| 2 1/4 | 7 1/8 | 8 | 186 | 37 | 9 |
| 2 | 6 1/4 | 6.30 | 140 | 28 | 8 |
| 1 7/8 | 5 3/4 | 5.55 | 127 | 25 | 8 |
| 1 3/4 | 5 1/2 | 4.85 | 112 | 22 | 7 |
| 1 5/8 | 5 | 4.15 | 94 | 19 | 6.5 |
| 1 1/2 | 4 3/4 | 3.55 | 82 | 16 | 6 |
| 1 3/8 | 4 1/4 | 3 | 72 | 14 | 5.5 |
| 1 1/4 | 4 | 2.45 | 58 | 12 | 5 |
| 1 1/8 | 3 1/2 | 2 | 47 | 9.5 | 4.5 |
| 1 | 3 | 1.58 | 38 | 7.6 | 4 |
| 7/8 | 2 3/4 | 1.20 | 29 | 5.8 | 3.5 |
| 3/4 | 2 1/4 | .89 | 23 | 4.6 | 3 |
| 5/8 | 2 | .62 | 15.5 | 3.1 | 2.5 |
| 9/16 | 1 3/4 | .50 | 12.3 | 2.4 | 2.25 |
| 1/2 | 1 1/2 | .39 | 10 | ? | 2 |
| 7/16 | 1 1/4 | .30 | 8 | 1.6 | 1.75 |
| 3/8 | 1 1/8 | .22 | 5.75 | 1.15 | 1.50 |
| 5/16 | 1 | .15 | 3.8 | .76 | 1.25 |
| 1/4 | 3/4 | .10 | 2.65 | .53 | 1 |

# STRENGTH OF WIRE ROPE

in tons of 2,000 pounds

### Wire Transmission Rope. One Hemp core surrounded by six strands of seven wires each.

| Diameter in Inches | Iron | Crucible Cast Steel | Extra Strong Crucible Cast Steel | Plow Steel |
|---|---|---|---|---|
| 2 ¾ | | | | |
| 2 ½ | | | | |
| 2 ¼ | | | | |
| 2 | | | | |
| 1 ¾ | | | | |
| 1 ⅝ | | | | |
| 1 ½ | 32 | 63 | 73 | 82 |
| 1 ⅜ | 28 | 53 | 63 | 72 |
| 1 ¼ | 23 | 46 | 54 | 60 |
| 1 ⅛ | 19 | 37 | 43 | 47 |
| 1 | 15 | 31 | 35 | 38 |
| ⅞ | 12 | 24 | 28 | 31 |
| ¾ | 8.8 | 18.6 | 21 | 23 |
| ⅝ | 6 | 13 | 14.5 | 16 |
| ⁹⁄₁₆ | 4.8 | 10 | 11 | 12 |
| ½ | 3.7 | 7.7 | 8.85 | 10 |
| ⁷⁄₁₆ | 2.6 | 5.5 | 6.25 | 7 |
| ⅜ | 2.2 | 4.6 | 5.25 | 5.9 |
| ⁵⁄₁₆ | 1.7 | 3.5 | 3.95 | 4.4 |
| ⁹⁄₃₂ | 1.2 | 2.5 | 2.95 | 3.4 |
| ¼ | | | | |

| Diameter in Inches | Wire Transmission Rope. One Hemp core surrounded by six strands of seven wires each. ||||
|---|---|---|---|---|
| | Iron | Crucible Cast Steel | Extra Strong Crucible Cast Steel | Plow Steel |

| Diameter in Inches | Wire Transmission Rope. One Hemp core surrounded by six strands of nineteen wires each. ||||
|---|---|---|---|---|
| | Iron | Crucible Cast Steel | Extra Strong Crucible Cast Steel | Plow Steel |
| 2 ¾ | 111 | 211 | 243 | 275 |
| 2 ½ | 92 | 170 | 200 | 229 |
| 2 ¼ | 72 | 133 | 160 | 186 |
| 2 | 55 | 106 | 123 | 140 |
| 1 ¾ | 44 | 85 | 99 | 112 |
| 1 ⅝ | 38 | 72 | 83 | 94 |
| 1 ½ | 33 | 64 | 73 | 82 |
| 1 ⅜ | 28 | 56 | 64 | 72 |
| 1 ¼ | 22.8 | 47 | 53 | 58 |
| 1 ⅛ | 18.6 | 38 | 43 | 47 |
| 1 | 14.5 | 30 | 34 | 38 |
| ⅞ | 11.8 | 23 | 26 | 29 |
| ¾ | 8.5 | 17.5 | 20.2 | 23 |
| ⅝ | 6 | 12.5 | 14 | 15.5 |
| 9/16 | 4.7 | 10 | 11.2 | 12.3 |
| ½ | 3.9 | 8.4 | 9.2 | 10 |
| 7/16 | 2.9 | 6.5 | 7.25 | 8 |
| ⅜ | 2.4 | 4.8 | 5.30 | 5.75 |
| 5/16 | 1.5 | 3.1 | 3.50 | 3.8 |
| 9/32 | | | | |

| Diameter in Inches | Wire Transmission Rope. One Hemp core surrounded by six strands of seven wires each. ||||
| --- | --- | --- | --- | --- |
| | Iron | Crucible Cast Steel | Extra Strong Crucible Cast Steel | Plow Steel |
| ¼ | 1.1 | 2.2 | 2.43 | 2.65 |

# SPLICING TRANSMISSION CABLES

Wherever wire rope transmissions are used it is necessary to splice the rope or cable so that it will run smoothly over the sheave wheels. For this purpose a long splice is invariably used. (Taken from "American Wire Rope" published by American Steel and Wire Company). The tools required are a small marlin-spike, nipping cutters, and either clamps or a small hemp rope sling with which to wrap around and untwist the rope. If a bench vise is accessible, it will be found very convenient for holding the rope.

"In splicing a rope, a certain length is used up in making the splice. An allowance of not less than 16 feet for ½ inch rope, and proportionately longer for larger sizes, must be added to the length of an endless rope, in ordering. The length of splice relation to the diameter of the rope is approximately 50:1."

This extra length is equal to the distance EE´ in Fig. 117. The additional length recommended for making a splice in different sizes of wire rope is as follows:

| Diam. of Rope in Inches | Extra Length Allowed for the Splice, Feet |
|---|---|
| 3/8 | 16 |
| 1/2 | 16 |
| 5/8 | 20 |
| 3/4 | 24 |
| 7/8 | 28 |
| 1 | 32 |
| 1 1/8 | 36 |
| 1 1/4 | 40 |
| 1 1/2 | 44 |

Fig. 117. Having measured carefully the length the rope should be after splicing and marked the points M and M´, unlay the strands from each end E and E´, to M and M´, and cut off the hemp center at M and M´.

Fig. 118. First. Interlock the six unlaid strands of each end alternately, cutting off the hemp centers at M and M´, and draw wire strands together, so that the points M and M´ meet, as shown.

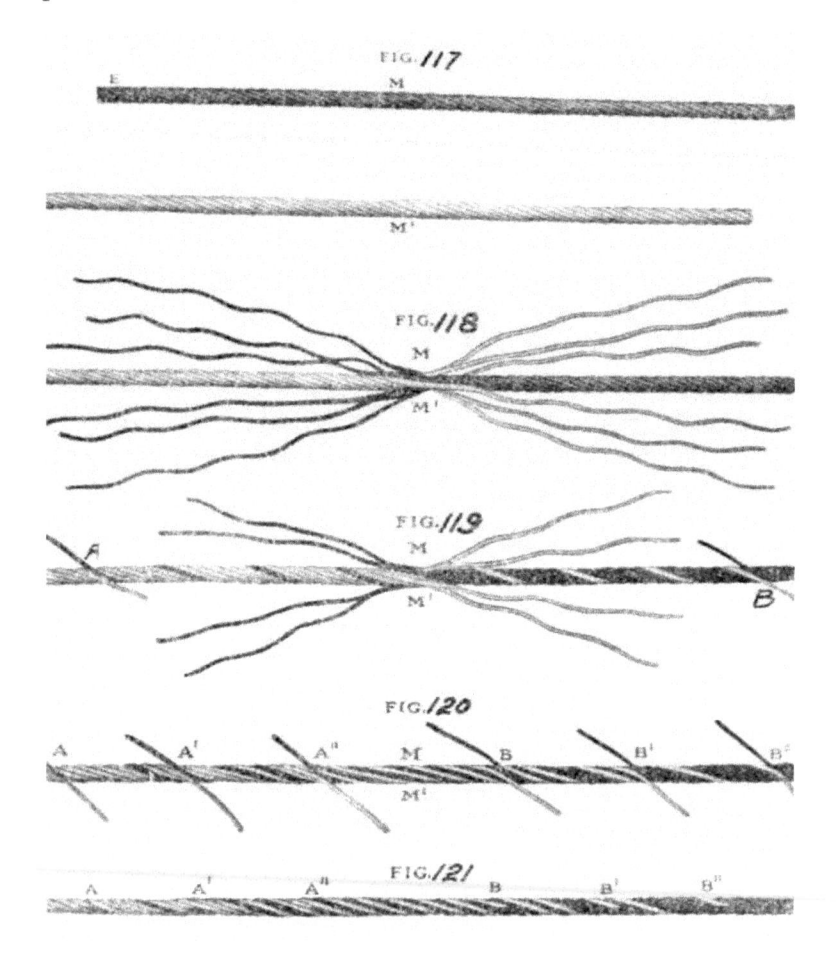

FIG. 117

FIG. 118

FIG. 119

FIG. 120

FIG. 121

Fig. 119. Second. Unlay a strand from one end, and following the unlay closely, lay into the seam or groove it opens, the strand opposite it belonging to the other end of the rope, until there remains a length of stand equal in inches to the length of splice EE´ in feet, e. g., the straight end of unlaid strand A on one-half inch rope equal 16 inches for 16 foot splice. Then cut the other strand to about the same length from the point of meeting, as shown at A.

Fig. 119. Third. Unlay the adjacent strand in the opposite direction, and following the unlay closely, lay in its place the corresponding opposite strand, cutting the ends as described before at B.

The four strands are now laid in place terminating at A and B, with eight remaining at M and M´ as shown in Fig. 119.

It will be well after laying each pair of strands to tie them temporarily at the points A and B.

Fig. 120. Pursue the same course with the remaining four pairs of opposite strands, stopping each pair of strands so as to divide the space between A and B into five equal parts, and cutting the ends as before.

All the strands are now laid in their proper places with their respective ends passing each other.

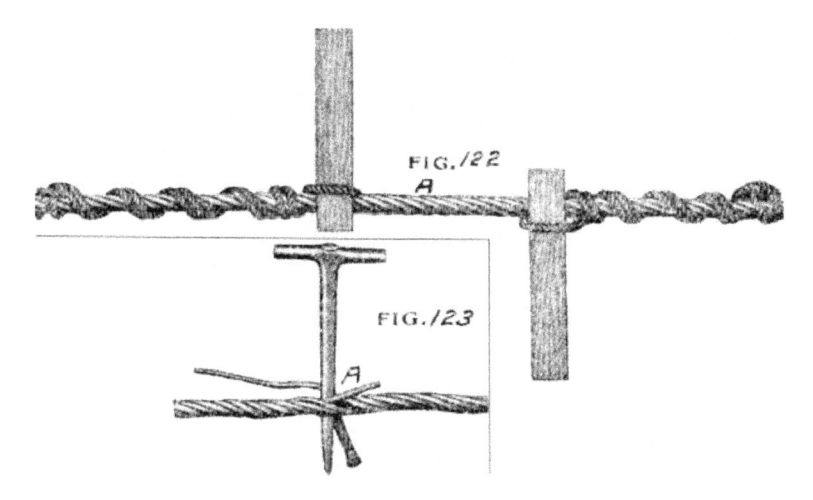

FIG. 122
A

FIG. 123
A

All methods of rope splicing are identical up to this point; their variety consists in the method of securing the ends.

Fig. 121. The completed splice with ends secured results in a cable with scarcely any enlargement at that point. A few days' use will make it difficult to discover at all.

The final part of the splice is made as follows:

"Clamp the rope either in a vise or with a hand clamp at a point to the left of A (Fig. 119), and by a hand clamp applied near the right of A open up the rope by untwisting sufficiently to cut the hemp core at A, and seizing it with nippers, let your assistant draw it out slowly. Then insert a marlin spike under the two nearest strands to open up the rope and starting the loose strand into the space left vacant by the hemp center, rotate the marlin spike so as to run the strand into the center. Cut the hemp core where the strand ends, and push the end of hemp back into its place. Remove the clamps and let the rope close together around it. Draw out the hemp core in the opposite direction and lay the other strand in the center of the rope in the same manner. Repeat the operation at the five remaining points, and hammer the rope lightly at the points where the ends pass each other at A´, B´, etc., with small wooden mallets, and the splice is complete, as shown in Fig. 121."

A rope spliced as above will be nearly as strong as the original rope, and smooth everywhere. After running a few days, the splice, if well made, cannot be pointed out except by the close examination of an expert.

Fig. 122. If a clamp and vice are not obtainable, two rope slings and short wooden levers may be used to untwist and open up the rope.

Fig. 123. A marlin spike is absolutely necessary in order to separate the strands in making a splice in steel cable.

## POWER TRANSMITTED BY WIRE ROPE

### Wire Rope Drives

| Diam. of Wheel in Feet | No. of Revolutions Per Minute | Diam. of Rope | Horse Power |
|---|---|---|---|
| 3 | 80 | 3/8 | 3 |
| 3 | 100 | 3/8 | 3 1/2 |
| 3 | 120 | 3/8 | 4 |
| 3 | 140 | 3/8 | 4 1/2 |
| 4 | 80 | 3/8 | 4 |
| 4 | 100 | 3/8 | 5 |
| 4 | 120 | 3/8 | 6 |
| 4 | 140 | 3/8 | 7 |
| 5 | 80 | 7/16 | 9 |
| 5 | 100 | 7/16 | 11 |
| 5 | 120 | 7/16 | 13 |
| 5 | 140 | 7/16 | 15 |
| 6 | 80 | 1/2 | 14 |
| 6 | 100 | 1/2 | 17 |
| 6 | 120 | 1/2 | 20 |
| 6 | 140 | 1/2 | 23 |
| 7 | 80 | 9/16 | 20 |

| Diam. of Wheel in Feet | No. of Revolutions Per Minute | Diam. of Rope | Horse Power |
|---|---|---|---|
| 7 | 100 | 9/16 | 25 |
| 7 | 120 | 9/16 | 30 |
| 7 | 140 | 9/16 | 35 |
| 8 | 80 | 5/8 | 26 |
| 8 | 100 | 5/8 | 32 |
| 8 | 120 | 5/8 | 39 |
| 8 | 140 | 5/8 | 45 |
|  |  | 9/16 | 47 |
| 9 | 80 | 5/8 | 48 |
|  |  | 9/16 | 58 |
| 9 | 100 | 5/8 | 60 |
|  |  | 9/16 | 69 |
| 9 | 120 | 5/8 | 73 |
|  |  | 9/16 | 82 |
| 9 | 140 | 5/8 | 84 |
| 10 | 80 | 5/8 | 64 |
|  |  | 11/16 | 68 |
| 10 | 100 | 5/8 | 80 |
|  |  | 11/16 | 85 |
| 10 | 120 | 5/8 | 96 |
|  |  | 11/16 | 102 |
| 10 | 140 | 5/8 | 112 |
|  |  | 11/16 | 119 |
| 12 | 80 | 11/16 | 93 |
|  |  | 3/4 | 99 |
| 12 | 100 | 11/16 | 116 |
|  |  | 3/4 | 124 |

| Diam. of Wheel in Feet | No. of Revolutions Per Minute | Diam. of Rope | Horse Power |
|---|---|---|---|
| 12 | 120 | 11/16 | 140 |
|  |  | 3/4 | 149 |
| 12 | 120 | 7/8 | 173 |
| 14 | 80 | 1 | 141 |
|  |  | 1 1/8 | 148 |
| 14 | 100 | 1 | 176 |
|  |  | 1 1/8 | 185 |

## MINIMUM DIAMETERS OF SHEAVES FOR POWER TRANSMISSION BY WIRE ROPES

(All Dimensions in Inches)

| Rope Diam. | Steel 7-Wire | Steel 19-Wire | Iron 7-Wire | Iron 19-Wire |
|---|---|---|---|---|
| 1/4 | 20 | 12 | 40 | 24 |
| 5/16 | 25 | 15 | 50 | 30 |
| 3/8 | 30 | 18 | 60 | 36 |
| 7/16 | 35 | 21 | 70 | 42 |
| 1/2 | 40 | 24 | 80 | 48 |
| 9/16 | 45 | 27 | 90 | 54 |
| 5/8 | 50 | 30 | 100 | 60 |
| 11/16 | 55 | 32 | 110 | 66 |
| 3/4 | 60 | 35 | 120 | 72 |
| 7/8 | 70 | 41 | 140 | 84 |
| 1 | 80 | 47 | 160 | 96 |
| 1 1/8 | 90 | 53 | 180 | 108 |
| 1 1/4 | 100 | 58 | 200 | 120 |
| 1 3/8 | 110 | 64 | 220 | 132 |

| Rope Diam. | Steel 7-Wire | Steel 19-Wire | Iron 7-Wire | Iron 19-Wire |
|---|---|---|---|---|
| 1 ½ | 120 | 70 | 240 | 144 |

## DIAMETER OF MINIMUM SHEAVES IN INCHES, CORRESPONDING TO A MAXIMUM SAFE WORKING TENSION.

| Diam. of Rope in Inches | Steel 7-Wire | Steel 12-Wire | Steel 19-Wire | Iron 7-Wire | Iron 12-Wire | Iron 19-Wire |
|---|---|---|---|---|---|---|
| ¼ | 19 | 15 | 11 | 39 | 31 | 23 |
| 5⁄16 | 24 | 19 | 14 | 49 | 38 | 29 |
| ⅜ | 29 | 22 | 17 | 59 | 46 | 35 |
| 7⁄16 | 34 | 26 | 19 | 69 | 54 | 41 |
| ½ | 38 | 30 | 22 | 79 | 61 | 47 |
| 9⁄16 | 43 | 33 | 25 | 89 | 69 | 52 |
| ⅝ | 48 | 37 | 28 | 99 | 77 | 58 |
| 11⁄16 | 53 | 41 | 31 | 109 | 84 | 64 |
| ¾ | 58 | 44 | 34 | 119 | 92 | 70 |
| ⅞ | 67 | 52 | 39 | 138 | 107 | 81 |
| 1 | 77 | 59 | 45 | 158 | 123 | 93 |